Globalization, Transition and Development in China

This book aims to explain China's development strategy and its underlying forces, and the success of this strategy. It examines China's gradualist approach, which emphasizes development first, and regards transition and globalization as secondary, enacting liberalization of domestic markets, and integration into the world economy in a paced way, avoiding dramatic changes which might impede or even reverse development, and argues that this approach is broadly correct. It considers China's failures, including the failure to build large globally competitive corporations despite the intention to do this, and shows how China's economic strategy has been implemented in detail with a case study of the large and important coal industry.

Huaichuan Rui received her PhD from the Judge Institute of Management, University of Cambridge, where she is currently a Research Fellow. Her main interests are in globalization and its impact on both developed and developing countries, and China's coal and related industries (steel, power and raw materials). She also teaches in Cambridge and is consultant to various investment companies.

RoutledgeCurzon Studies on the Chinese Economy

Series editors

Peter Nolan
University of Cambridge
Dong Fureng
Beijing Univerity

The aim of this series is to publish original, high-quality, research-level work by both new and established scholars in the West and the East, on all aspects of the Chinese economy, including studies of business and economic history.

Globalization, Transition and Development in China

The case of the coal industry

Huaichuan Rui

Routledge
Taylor & Francis Group

LONDON AND NEW YORK

First published 2005
by Routledge
2 Park Square, Milton Park, Abingdon, Oxon OX14 4RN

Simultaneously published in the USA and Canada
by Routledge

711 Third Avenue, New York, NY 10017

Routledge is an imprint of the Taylor & Francis Group

First issued in paperback 2012

© 2005 Huaichuan Rui

Typeset in Times by
HWA Text and Data Management, Tunbridge Wells

British Library Cataloguing in Publication Data
A catalogue record for this book is available from the British Library

Library of Congress Cataloging in Publication Data
A catalog record for this book has been requested

ISBN 978-0-415-33319-1 (Hardback)
ISBN 978-0-415-65499-9 (Paperback)

In memory of my parents, Rui Qiansheng and Zhang Shuzhen

Contents

Figures

Tables

Foreword

The Chinese coal industry sounds a boring topic. In fact few topics touch so closely the heart of China's political economy in the early twenty-first century. Indeed, the coal industry is of central importance in global development in the early twenty-first century. In both China and the USA, as well as in many smaller countries, coal is the most important source of primary energy for electricity generation. Managing the environmental consequences of the huge generation of carbon dioxide that results from burning coal in power stations is a central issue for the world to tackle in the years ahead. However, in the absence of severe penalties for using coal in power stations, coal remains a highly competitive source of primary energy. Coal is not only still extremely important in China, but its coal output is rising fast, and shows every sign of continuing to rise steadily. In the absence of countervailing measures, it is likely that China's already huge coal output will rise to levels far above those of today.

Huaichuan Rui's book is based on unique fieldwork in three different types of Chinese mining enterprise – township and village enterprises, old state-owned enterprises and new modern coal companies. She also places her analysis in the context of global change in this industry and its implications for China. Through the study of this dirty, 'boring' industry, Rui has provided rich insights into the policy challenges that confront the Chinese government today.

Despite China's high-speed growth, and the fears of Western countries, especially the USA, that they will soon be 'overtaken' by China, the country remains extremely poor, facing massive development challenges. There are still several million people working in China's small, local coalmines. Their lives are little different from those of miners in Britain during the Industrial Revolution. However, they remain highly competitive in domestic markets, due to low labour and capital costs. They typically also have low transport costs, supplying predominantly local needs for low-priced unwashed, ungraded coal. Alongside China's modernization, output and employment has grown rapidly in this sector, despite government attempts to control its growth. Conditions of work are extremely dangerous in terms both of short-term risks of accidents, and long-term risks to miners' health. The sector also is extremely wasteful due to the low extraction rate of mine reserves, and damage to the natural environment due to the primitive conditions of coal transportation from such mines. Areas with large numbers of small coalmines are deeply polluted. However, the sector offers desperately need employment for poor people. The task of devising the 'correct' development policy for this sector is a great challenge. There are no easy answers.

The coal sector's numerous large state-owned mines face another type of challenge. The core mines in the sector began production many years ago, having been established on the richest, most easily located coal seams. Today, many of these giant mines, with a huge workforce, face a bleak future, having exhausted the most easily worked deposits, and facing exhaustion of reserves and rising costs of extraction. The sector is heavily loss-making, and massively indebted to the state banks. Mines are typically located in relatively remote areas in which the alternative employment opportunities for miners are slim. Miners in all countries are notoriously prone to take political action in defence of their interests. However, the Chinese government cannot indefinitely prop up the huge ailing state-owned mines with loans from the state-owned banks. In this respect, the mining sector illuminates some of the deepest problems of the transitional economies. This sector also presents deep policy challenges for the Chinese government.

China's declining state-owned coal mines face a severe challenge from giant, modern coal mines, both domestic and, increasingly, foreign. On the one hand, the Chinese government must consider the difficult issue of how to prevent widespread unemployment and social disorder in the mining industry. On the other hand, it must consider the intense business challenge to its indigenous mines from the global giant mining companies, which have consolidated their position since the 1980s. A small group of global giants, such as Rio Tinto and BHP Billiton, has emerged through a series of mergers and acquisitions, to comprehensively dominate the international coal trade. These giant firms benefit from economies of scale in the acqusition and development of mining sites, in the purchase and operation of giant pieces of capital equipment, in the construction of a global brand, in the employment of the world's best human resources in this sector, and in the application of advanced techniques of 'system integration' across the whole value chain of the mining industry, from exploration to the on-time final delivery of washed and precisely graded coal to global customers.

In order to meet this challenge, the Chinese government has attempted to nurture a small group of giant indigenous mining firms that can compete with the global giants of the industry. The most famous of these is the Shenhua Coal Group, which Dr Rui's research examines closely. Shenhua is a large, modern coal company, with huge deposits of high-quality coal and a small workforce that enables it to keep its costs down. It also has its own dedicated railway line several hundred kilometres in length and dedicated port facilities. It is attempting to replicate the structure of the global giants in many key respects. It is beginning to present a formidable competitive challenge not only for the domestic, declining state-owned firms, but also for the global giants. It has already begun to compete successfully in coal markets in north-east Asia. However, as Dr Rui's research shows, emerging giant firms such as Shenhua still remain heavily constrained by the nature of the political environment in which they operate. Just as it was about to launch into accelerated domestic and international expansion, Shenhua was forced to accept the take-over of the 'Five Western District Mines', which employed many times the workforce of Shenhua and overall were loss-making. This was extremely difficult for Shenhua's management to deal with. Trying to build a modern competitive coal company in such an uncertain institutional environment is a massive management challenge.

Dr Rui's study provides a rare integration of in-depth case studies of individual Chinese firms of different types with a deep awareness of the domestic macro-economic environment in which they operate, an understanding of the competitive challenges they face from each other as well as, increasingly, from the forces of global competition. It is unique in its close integration of the threefold challenges that face China's policy-makers today: development, transition and globalization.

Peter Nolan
University of Cambridge

Acknowledgements

My deepest appreciation goes to Professor Peter Nolan. Without his 'Chinese Big Business Programme', I would never have become involved in researching China's industry in general and the coal industry in particular; without his every encouragement, it would have been much more arduous for me to cope with so many difficulties during the time of preparing this book; and most importantly, without his unique love of China that combines both sharp criticism and fundamental support, I could not have had such a profound understanding of the difficulties facing the serious scholar and researcher, and his or her responsibility to remain fully aware of these difficulties, particularly in an often prejudiced academic environment.

Professor Geoff Meeks and Professor John Sender made extremely valuable comments on the original draft of this book, for which I am deeply grateful.

I am indebted to so many Chinese companies, government departments and agencies for their considerable support of the fieldwork I carried out in China between 1999 and 2003 by accepting interviews, providing detailed data and offering patient discussions that often contributed fresh opinions on the subjects raised. My special thanks go to the Shenhua Group and its branches, and the Jixi Mining Bureau. Thanks also go to the State Development and Reform Commission, Qitaihe Clean Coal Corporation; several TVE coalmines; the State Coal Administration (current National Coal Association); the Shanxi Coal Administration; the Datong Coal Administration; the Shenmu County government, the Shenmu Coal Administration; the Ministry of Railway; the State Power Corporation and the Tianjin Port Authority.

Numerous individuals have provided considerable help in my fieldwork in China. I wish especially to thank the following: Professors Lin Kaiyuan and Wang Wenfei of China's Coal Consultancy and Mr Duan Zhijie of the State Development Bank, who not only initiated my research on coal but also continued to support this research in every possible way. Mr Ye Qing, former Deputy Director General for the State Planning Commission and current Chairman of Shenhua, who not only supported my case study on Shenhua but also was unsparing in giving his precious time to my two long interviews presenting me with a comprehensive overview of the development both of the coal history of Shenhua in particular and of the whole of China in general.

Also the following members of the Shenhua Group: Messrs Wang Xiaolin, Zhou Dayu, Zhang Yuzhuo, Dai Shaocheng, Yang Jingcai, Xie Youquan, Wei Shenghong, Ma Zhifu, Deng Yu, Han Jianguo, Wang An, Kou Ping, Jia Yuzhu, Shan Bingjun, Ma Jun, Yang Wangli, Yan Penggui, Li Song, and Ms Wu Yan. Of the Jixi Mining Bureau the former Director Ren Zhenming and Deputy Directors Li Guozhong, Li Naidong, Hao Fukun, and the much regretted Mr. Liu Dingli who was killed in the accident in Jixi 6.20 and to whose family I offer my sincere condolences. The three *Records of the Jixi Mining Bureau* he lent to me are still on my bookshelf! These men provided me every possible assistance in my fieldwork during the whole month I stayed in Jixi. My sincere thanks too to Mr Lu Yansun, the former Deputy Minster of Machinery Industry and current Deputy Director of China Machinery Industry Association; Messrs Pan Jiazhen and Xie Juchen of the State Power Corporation, Mr Jin Langchuan of Huaneng Group, and Mr. Wei Pengyuan of the State Development and Reform Commission. Also to all my friends in the Information Center of the Ministry of Water Resource, especially Messrs Cai Yang, Chang Zhihua, Cheng Yilian, Zhou Weixu and their families. To the Directors of Shanxi Coal Administration, Mr. Du Fuxin and Gong Anku, and those who did not want their companies to be named including Kang Zhuang, Luo Zhilan, Xu Gaolin and Liu Minfang. Finally, I would give my sincere respect to all Chinese coal workers, whose life and spirit have taught me so much.

Back in Cambridge, there are many of my friends and colleagues who have unstintingly assisted me with their generous support. I am considerably indebted to Mr Alan Shipman, whose profound knowledge of economics, energy resources, transition and globalization issues has enabled me to add so much valued depth to this book. Moreover, his kindness, encouragement and friendship are of a quality I have come to appreciate enormously. I am also indebted to Dr Zhang Mei, whose detailed guidance in every procedure of my research has always been most helpful. Similarly to Mr David Weston, Dr Dylan Sutherland and Ms Jane Cowan who have been so helpful in many different ways.

I started my PhD study in Cambridge, on which this book mainly bases, two months after having given birth to my daughter in Australia. Without the great encouragement and support from my husband Junming, who took on all the responsibility to look after our son Dai Rui and baby Yale, I could not have started this research. Moreover, without his generous financial support sacrificing the savings from many years hard work, my studies would have been impossible. I shall never be able to return this love from my family in full.

Finally, my understanding of China, of life, and of human weaknesses and virtues, is rooted in the education I received from my parents and my close observation of the hardships they endured throughout much of their lives. It is they who made me who I am, and I regret that I never fully acknowledged this to them when they were still alive. It is to their memory therefore that I dedicate this book.

Abbreviations

CBE	Community and Brigade Enterprise
CBM	coalbed methane
CCC	China Coal Consultancy
CCIY	*China's Coal Industry Yearbook*
CCP	Chinese Communist Party
CCT	clean coal technology
CERC	China Energy Research Committee
CNCIEC	China National Coal Import and Export Corporation
COE	collective owned enterprises
CRS	contract responsibility system
CSD	China Statistic Digest
DTI	Department of Trade and Industry, United Kingdom
EIA	Energy Information Administration, United States
FDI	foreign direct investment
FT	*Financial Times*
FWB	Five Western Bureaux
GAP	general allocation plan
GDP	gross domestic product
HRS	household responsibility system
IEA	International Energy Agency
IMF	International Monetary Fund
km	kilometre
LDC	less developed country
M&A	merger and acquisition
MCI	Ministry of Coal Industry
MNC	multi-national companies
MSDW	Morgan Stanley Dean Witter
Mt	million tons
Mtoc	million tons of coal equivalent
NIE	newly industrial enonomy
PRC	People's Republic of China
R&D	research and development
SAMC	State Asset Management Commission, successor of SETC

SC	State Council
SCA	State Coal Administration
SDB	State Development Bank
SDPC	State Development Planning Commission, successor of SPC
SDRC	State Development Reform Commission, successor of SDPC
SETC	State Economic Trade Commission
SOEs	state-owned enterprises
SPC	State Planning Commission
SSB	State Statistical Bureau
TVEs	township- and village-owned enterprises
TVGs	township and village governments
UNCTD	United Nations Conference on Trade and Development
WTO	World Trade Organization

Introduction

Motivation

On a cold and rainy winter day in Cambridge in 1999 an international seminar was discussing an equally 'cold' topic: China's coal industry. A case study of the best coal company – the Shenhua Group – was presented and some of the world's leading coal, oil and gas companies, as well as some banks, expressed considerable interest in the case. Shenhua is a real superstar in China's coal industry. It has the best and largest coalfield, a dedicated rail link and port, the most advanced imported equipment, global levels of productivity and the government's comprehensive support. However, most of the seminar attendees were surprised to hear that in 1998 the Chinese government had forced this superstar to 'merge' with five traditional state-owned coalmines, most of which were in great financial difficulty. This so-called 'forced marriage' changed Shenhua's basic features at a stroke. After the seminar most of the delegates felt confused by this enforced change in Shenhua, asking themselves why the government would do such a foolish thing.

With this question in mind, I interviewed the Chinese delegates immediately after the seminar. Some common points were emphasized by all. China is the world's largest in both coal production and consumption. Coal provides about 70 per cent of the country's primary energy source, generates over 80 per cent of the electricity and supports the world's largest steel production. As well as this, the coal industry employed over eight million employees before the large-scale closure of township- and village-owned (TVE) coalmines. However, it was also one of the most problematic industries: there was a market surplus; it was the leading loss-maker of all industries in China and had had over three million redundancies; it suffered from huge financial difficulties; it had high accident rates; it suffered frequent demonstrations by coal workers; and finally, it was regulated by numerous unreasonable government policies.

Yet, despite these problems and the importance of the industry, almost no comprehensive research has yet been done on this industry in the period of reform since 1978. It was therefore impossible to get a comprehensive and unbiased answer to the question raised at the seminar.

Motivated by a desire to fill these *lacunae*, as well as being moved by the plight of impoverished coal workers, I started my research on the coal industry and have continued to do so since 1999. After visiting numerous coal companies and mines, interviewing hundreds of people, consulting various government policy makers and coal-related industrial departments, and observing coal workers' activity underground, I gradually realized that the coal industry was one of the best cases from which to understand the country's past, present and future developmental strategy as a whole.

Considering its strategic importance, the government had a tight grip on the industry under the command economy, by controlling the investment, output, price, transportation and wage structure of state-owned coalmines. When reforms started in 1978, the coal industry was dominated by state-owned coal companies (SOEs), which included 104 key SOE coal bureaux and less than 2,000 local SOE coal producers, producing over 85 per cent of the national output. The remaining 15 per cent was produced by commune and brigade (CBE) coalmines, which were officially re-named as township- and village-owned (TVE) coalmines in 1984.[1]

However, as in most former command economies, shortages of most commodities were a common phenomenon. To meet higher demand for coal caused by the beginnings of the economic reforms, from the mid-1980s the central government strongly encouraged the development of the TVE coalmines. By 1996 their national market share rapidly increased from 15 to 50 per cent. This huge increase in output ended the history, decades long, of coal shortage and even turned the coal market to surplus by the mid-1990s.

Currently China's coal industry still consists of these three categories of producers:

1   key SOE coal bureaux (companies), which previously belonged to the Ministry of Coal Industry, but were later in 1998 decentralized to local provinces and autonomous regions;
2   local SOE mines (companies), which are operated by either local provincial, prefecture or county governments; and
3   TVE mines (companies) which, in fact, include those under private ownership.

Huge problems exist in this industry. One of the most prominent is diseconomies of small scale. Among 50,000 coal companies in 1999, 'none of them has a significant market share ... and over-competition has caused excessively scattered capital and technical investment and lack of economies of scale' (Yan *et al.* 2000).

Closely connected to the problem with these diseconomies of scale is that the coal industry has been in over-supply since the mid 1990s because of the rapid unregulated free entry of small TVE coalmines into the market. Failing to compete with the TVE's low cost of coal, most SOE coal bureaux had to reduce their price and suspend full use of their capacity. Making losses became unavoidable. However, they still had to keep functioning and fulfilling their social welfare obligations. Because of these commitments the coal industry has been one of the largest loss-makers for many years. Facing SOEs' worsening losses, and also being aware of TVEs' numerous problems (such as the worst records in the world for safety and environmental protection), the government has, since 1998, attempted to close most TVE coalmines.

While confronting problems of both SOEs and TVEs, China also faces challenges from economic globalization in general, meeting the requirements as a new member of the World Trade Organization (WTO) in particular. To prepare for this, China is actively building globally competitive indigenous coal corporations. Shenhua Group has been built in this spirit, and it has already emerged as one of the world's most competitive companies. However, as noted, the very same government, in 1998, also forced a 'merger' with a number of traditional large loss-making SOE coal bureaux because it had failed to decentralize these loss-makers to local governments.

The government was criticized from all sides as a result of these measures. SOE coal producers complained that the government should have closed TVEs earlier because of their poor record on safety and environmental protection. If TVE coalmines had been successfully closed, SOEs would not have been in such difficulties because coal would not have been in such surplus, which caused dropping coal prices (before 1980s SOEs enjoyed monopoly rents). Having been faced by the fact that coal was already in surplus, and that they were neither able to exploit their full capacity, nor sell their coal or make adequate investments, these SOEs challenged the government, demanding to know why huge sums were being invested to build new capacity, such as that found at Shenhua. Would this, they wanted to know, not result in an even higher output and in turn more surplus and more redundancies?

TVE coalmines insisted that the government should not simply close them. They were initially encouraged to produce more when coal was in shortage and they frequently contributed 50 to 95 per cent of local revenues and were a major source of rural development. If all the TVE coalmines were closed, could the local economy maintain its public spending and the living standards of local people? Meanwhile, TVEs believed that SOE losses were not entirely a result of competition from TVEs, but instead it resulted from the SOEs' lack of incentives to improve their management and efficiency. Why, in this case, should the government keep uncompetitive SOEs alive but force competitive TVE coalmines to close? Finally, should the many legally recognized TVE coalmines also be closed without any compensation at all?

Moreover, the large coal groups, such as Shenhua, were also unable to understand why the government wanted them to reinforce their competitive capacity on the global playing-field on one hand, while on the other force them to merge with large loss-making SOE coalmines, thus obviously constraining their ability to compete with the global giants.

Policymakers in China, then, suffered a range of challenges which can be placed in three categories. First, they faced the challenges arising from the basic difficulties of *development*. Second, they faced challenges arising from those of the *transition* from central planning. Third, they faced challenges arising from the pressures of *globalization*. Very explicitly in the coal industry, the challenge from development derived from the fact that in spite of problems, the development of TVE coalmines was unstoppable in the face of pressure from the rural population to climb out of poverty and the national demand for sufficient fuel. The challenge from transition centred on the need to transform the large loss-making SOE coal bureaux and the extreme difficulties in accomplishing this. The challenge from globalization was reflected in the wider gap between China's coal companies and their giant international counterparts, and the urgent need to build large indigenous globally competitive firms to pursue internationally competitive capacity. However, this is extremely problematic because TVE coalmines are difficult to close and SOE coalmines have to keep functioning.

These challenges of *development, transition* and *globalization* on the one hand can be seen to complement one another while at the same time generating various conflicts. The complexity of their inter-relationship and the difficulties the Chinese government faces in dealing with them is a classic example which can help us more clearly to understand China's development strategy and its underlying forces.

Since 1978 China has recognized that 'development is the paramount force' although this exact slogan was not officially formulated by Deng Xiaoping until 1992. To promote development China has undertaken a gradualist internal transition and also cautiously opened its doors to international trade and foreign investment. After two decades her economic success has become widely recognized. However, China is a vast country and its overall level of development, particularly in rural and western regions, is still low. The gradualism of transition has been consistently criticized in the light of evidence from the large loss-making and unprivatized SOEs. The country has also failed to build globally competitive corporations on a large scale, even though it took up this challenge comparatively early in its reforms with an explicit industrial policy.

Kornai (2000) concedes that few scholars recognized China's reform even in the early 1990s. Although a decade later more people now accept that gradualism is superior to shock therapy (including even Kornai), this current belief in gradualism is mainly based on a superficial observation of China's good economic performance and rarely on a deep understanding of the necessity of deploying such a strategy. This is why, until very recently, some scholars were still wondering whether fundamental flaws may exist in the Chinese approach, because they supposed that the high GDP growth might be overstated.

An important, though broad, question thus arises of how one might best understand China's development strategy and its underlying forces. This question can in fact be divided into a number of smaller questions. How to explain both the successes and failures? Why should the economic development be placed as a top priority and non-SOEs be encouraged to develop? Why should a gradualist strategy be deployed and should it be pursued in the future? Why should globally competitive, indigenous firms be built through industrial policy? Why does the government still intervene in economic affairs so persistently and should it still be the case in the future? The answers to such questions are extremely important for development strategy in China in the twenty-first century. They are also important for developing countries as a whole, many of which face similar problems, and important as a guide to new thinking in development economics.

As noted, based on the case studies of the Chinese coal industry, this research finds the underlying answers in the three key interlinked challenges from development, transition and globalization. The three reinforce one another while at the same time also disrupting one another.

Transition and integration into the world economy, for example, may assist development, but too dramatic a transition and liberalization of the domestic market to external market forces may impede or even reverse economic development. China has succeeded because it understood that development is a paramount challenge and an ultimate goal, so transition and integration with the world must be taken cautiously to secure, rather than to radically disturb, economic development. However, as all three challenges are so overwhelming, to pursue one may mean having to delay or even sacrifice another. The best solution, therefore, requires a complex mix of considerations in respect to these three challenges. This is why China has enjoyed qualified success, succeeding in some areas and failing in others. China's development strategy, though imperfect, has proved practical and effective in the often difficult conditions created by intertwined forces of development, transition and globalization pressures.

This study

The purpose of this study is to provide a better understanding of China's developmental strategy over the past two decades or more based on case studies of China's coal industry. It examines how the three parallel challenges from *development, transition* and *globalization* determine that reform must be handled cautiously, experimentally, innovatively and in a balanced way, with the state playing a significant and irreplaceable role.

Although there is a large amount of research on development, transition and globalization, few studies, if any, associate these three themes in a single work. Research on the relationship between transition and development, or between globalization and development, exists. None, however, examines how the three conflict while at the same time assist each other, or how this relationship has made the state's role particularly necessary. This study hopes to fill this vacuum as well as provide a description of China's coal industry, another area seldom researched.

The three subjects are quite broad. It is necessary, therefore, to narrow and define their context for this study. There are numerous definitions and understanding on 'development' (Clark 1999: 31–2), but this study focuses on economic development, an improvement of people's living standards, and a reduction in absolute poverty, unemployment and income inequality (Dudley Seers's criteria).

In contrast to the concept of development, transition from the command economy is a relatively new concept. A recent study on transition captures the general attitude towards this concept. It defines it as 'a process involving a fundamental shift leading from the late socialist centrally-planned economy based on the dominance of state ownership toward a free market, with the private sector in the key role' (Kolodko 2000: 1). This definition, however, is not really adequate. China has never claimed that it will transfer from the current system to a purely free market system. In practice, as many scholars point out, what China has been pursuing seems neither socialism nor capitalism, neither command economy, nor free market economy. There is no guarantee, moreover, that private enterprises will fulfil the 'key role' in China's economy. In many senses the existing state sector, increasingly found in large-scale upstream operations, is fulfilling a key role. One thing all transition countries do have in common, including China, is that they definitely want to move away from the former command system. It seems, therefore, that transitional economies have left the position they were in before, but do not necessarily know where they are going. In China's case, *development* is the destination. Whatever the transition, if it brings China prosperity and stability, it will be a good transition. In this sense, transition is a non-linear and path-dependent choice for different countries which are transforming from the command economy.[2]

Globalization, according to Held *et al.* (1999: 1), reflects

> a widespread perception that the world is rapidly being moulded into a shared social space by economic and technological forces and that developments in one region of the world can have profound consequences for the life chances of individuals or communities on the other side of the globe.

While this broad definition is perfectly acceptable, this book will focus on global economic and financial integration in particular.

Structure of this book

Chapter 1 reviews both the theoretical and empirical backgrounds relevant to China's three challenges as well as their inter-relationship and the role of the state. Chapter 2 makes a comprehensive examination of the three challenges as they are reflected in the coal industry, so that these challenges facing China can be illustrated in more detail. The following three chapters are the core chapters of this study, each of which examines a specific challenge. Chapter 3, based on the case study of China's TVE coalmines, demonstrates why and how the challenge of development is consistently the main challenge. Chapter 4, based on the case study of China's traditional SOE coal companies, shows why and how transition becomes an overwhelming challenge. Chapter 5, based on the case study of the Shenhua Group – China's government-created globally competitively indigenous coal group – explains why and how globalization is also an overwhelming challenge. Finally, Chapter 6 examines the inter-relationship between the three challenges and the role that the state has played in the past, and that it should play in the future, to deal with these challenges and their complicated interrelationships.

Methodology

Most methodology courses recommend students to start research from a basic hypothesis which is then investigated, using the investigation as a test of the hypothesis. This study has a different approach. There was no specific hypothesis when starting the investigation. After the preliminary findings of my investigation I soon realized that the case of the coal industry was a perfect illustration of China's development strategies. I maintained, however, my exploratory approach and remained open-minded, seeking truth from practice. This study, therefore, draws heavily on several periods of fieldwork in China since 1999, covering case studies of three major categories of coal producers.

Jixi Mining Bureau was first selected as the representative of SOE coal companies to examine its dramatic changes during the transition. Jixi used to be one of the largest and best coal bureaux under the command economy. After the mid-1990s, however, it became one of those mines in most difficulty. Since it was my first case study and I had few ideas about Jixi, or the industry as a whole, my purpose was to get as broad a background picture as possible, not only by gaining information of the Jixi Bureau, but also of their competitors and of the policy-making regime. Therefore, my investigation of Jixi was essentially an open-ended question, but one of some interest: why had Jixi turned from a successful mine to become one of the worst?

My investigations of Jixi started exactly at the beginning of the new millennium on 1 January 2000. After a warmly celebrated millennium party in Beijing, I took the train to Jixi in the far north-east of China, which was suffering huge losses and was, basically, bankrupt. Jixi, it seemed, was another world altogether (see Figure 4.7). I spent the whole of January in this small coal city, interviewing people in temperatures of –40° centigrade and knee-deep snow, observing redundant workers who stood in the roadside snow waiting for casual jobs. I felt deep sympathy with these 'coal people' and resentment towards government policies.

My study of TVE coalmines included those in Heilongjiang, Shanxi and Shaanxi provinces, and the Inner Mongolia Autonomous Region, balancing poor and good TVE coalmines in terms of their technical competence, management and levels of contribution to the local economy. Since the government was implementing a closure policy at exactly the same time as I was conducting my fieldwork, my focus was to examine the history of TVE coalmines from the previously dramatic rise starting in the mid-1980s to an equally dramatic fall from mid-1990s, and to understand the attitudes of TVE coalmines and local governments to poverty, to the coal industry, and to government's policies on TVE coalmines.

The simplest of mine-shafts, out-dated equipment, hard-working peasant workers, and horrible working conditions were the best indicators of the competitive capacity of TVE mines (see Figure 3.4). SOE coal workers were openly hostile to TVE mines because they had lost market share to these TVEs. But as one TVE miner noted, putting up a defence: 'Can you see an SOE miner working underground the whole day as I do?' In SOE coal bureaux the workers are not willing to work underground, so they have to hire peasants to do the most dangerous and heavy work. This is partly why SOEs are overstaffed and have very high wage and welfare costs. It was indeed sad that these redundant workers from SOEs had to stand in the snow waiting for jobs, but they still received their 218 yuan living allowance each month, and probably more, as the result of showing their anger through frequently organized demonstrations. However, TVE coal workers accepted the fact that there was no alternative to their doing these dangerous jobs. Their competitive advantage was accounted for by the low value of their lives.

After this research, I visited the Shenhua Group to examine why China needs, and whether she can build, globally competitive coal corporations with the support of the government industrial policy, but with fierce global competition and constraints from SOEs and TVEs. I visited Shenhua's Beijing headquarters on numerous occasions and interviewed its director Ye Qing and over half of the department directors. I also visited its coalfield and coalmines on the border between Inner Mongolia and Shaanxi province and its leased TVE coalmines in Shanxi province (see Figures 5.4 and 6.4). Seeing the most advanced equipment imported from abroad working in Shenhua's mines, noting their high productivity record, and looking at Shenhua's effective management and highly motivated employees, I was excited and confident: there appeared to be the hope for China's coal industry.

While the government is committed to Shenhua as the future of China's coal industry and to the restructuring of SOEs and TVEs, it must simultaneously be aware of the pressures from rural development, urban unemployment and social instability. The government has to play an important role in dealing with such conflicts of interest. I therefore also visited Datong City Coal Administration, Shanxi Provincial Coal Administration, the State Coal Administration and the State Economic Trade Commission to understand policy-making procedures and administrative structures in the coal industry as well as their effects on the development of the industry. In addition I also visited the Tianjin Port Authority, the State Power Corporation of China, and the Ministry of Railways, which are all closely involved in coordinating the coal industry and therefore very helpful for understanding the coal industry.

Although this book drew heavily from the major fieldwork carried out from 1999 to 2001 for my doctoral study, it has been considerably updated and improved as the result of my fieldwork in 2002 and 2003 as a research associate at the University of Cambridge. In fieldwork from June to July 2003 I interviewed Ye Qing again and visited the building site of Shenhua's recently started – China's first ever – coal liquefaction project. As it is also located in the Shenfu Dongshen coalfield, I derived much satisfaction and pleasure in being able to witness the recent achievement of the Shenhua Group.

A note on terminology

The term 'cadre' was used widely under China's command economy since the 1950s to refer to both government officials and enterprise managers. It is still in use now, although more people refer to government cadres as 'officials' and firm cadres as 'managers' including chief executive, president, vice president, manager and so on.

From 1949 to 1998 there were 104 key state-owned coal companies in China entitled 'bureaux'. Since 1998 when the Ministry of Coal Industry was abolished and the 104 coal bureaux were decentralized and subject to the administration of local governments, some have changed their titles to be limited companies or shareholding companies while some still keep the titles of bureau.

All '$' in this book refers to the United States dollar which is equivalent to 8.27 RMB yuan at the time of writing (April 2004).

All the 'tons' used in this book refer to UK measurement, which is equivalent to 2420lb.

1   The challenges facing China

> Development is a task of prime importance for the CCP to execute power and
> develop the country's prosperity.
>
> The 16th CCP Congress Report

Introduction

A challenge, by its nature, is something that one must confront. Reviewing China's
history of development during the last two decades and looking forward to the
future, it is clear that numerous issues have to be dealt with. The most important
of the challenges that China has to confront can be broadly thought of in three
different categories. First, the challenges that arise from development – from the
current state of relative underdevelopment. Second, those from transition – moving
from the legacy of the previous command economy to a market economy with
Chinese characteristics. Third, those from globalization – from the closed door to
close integration with the world economy.

By way of introduction, the next three sections of this chapter will examine
why these three challenges are the most important ones that China must confront.
The final section will then examine what the inter-relationship between these three
challenges is and why the state still has a crucial role in dealing with them.

Development

'Development' can be understood as a concept very broadly. In his dissertation,
Conceptualising Development, Clark (1999: 31–2) listed at least 30 abstract
concepts of development employed in economics and the social sciences.
Development in this study, however, mainly focuses on 'economic development',
simply thought of as an improvement of people's living standard, which is generally
accompanied by a reduction in absolute poverty.

Over the years the Chinese people have paid a high price in realizing that
development is a supreme challenge, and that it must be striven for, but in an
acceptable way. When the Chinese Communist Party (CCP) took over power in
1949, Chairman Mao commented that old China was 'poor and blank' (*yiqiong
erbai*). The country was in a bad state after years of civil war, foreign intervention

and invasion. Mao, like nearly all of the Chinese, had a very strong ambition to turn China into a wealthy and powerful state. With the background of the 'cold war' and central planning based on Soviet experience, Mao soon concentrated on developing heavy industry which tended to eclipse the importance attached to light industry and agriculture. Although some dramatic progress was undoubtedly made under Mao's regime (see, for example, Bramall 1989, 1993), his policy as a whole failed to improve people's living standards to the level originally hoped for, especially for the rural population (Breth 1977; Perkins 1994; Yao 2000).[1] And undermining some of the positive steps forward, Mao encouraged the nation's population growth[2], and this now fundamentally constrains China's development (Banister 1987). Mao was also a believer in 'big bangs'. During his regime 'mass movements' were frequently initiated. Some of these, such as the Great Leap Forward and the Cultural Revolution, turned out to be unprecedented human tragedies.

Mao's achievements and thinking on China's development led to disillusionment with Chinese socialism. Socialism was described by Marx and proclaimed in Mao's time as a brilliant egalitarian society where people were all equal (with access according to need to sufficient food, materials, education and public health services). In reality, under Mao's regime there was overwhelming poverty (especially in rural China). Perversely, inequality, particularly between the urban and rural areas, also existed. Disillusionment with socialism and the sincere wish to make life better created the mandate for Deng Xiaoping to shift China's development strategy, to transform its socialist command system, and to focus on economic growth and improving people's living standard. In Deng's (1992: 370) words, 'China is such an immense country. Without reform and opening the door, without economic development, without improving people's life, China could only expect death'. Therefore, 'only development is the paramount force' (*fazhan caishi ying daoli*) (Deng 1992: 377). Reforms since the 1980s have broadly attempted to follow this line of thinking. As 80 per cent of the Chinese are basically peasants, to promote development mainly means promoting the living standards of this group.

Development is a supreme challenge in China basically because of its underdevelopment. An underdeveloped country is identified as having 'low per capita GDP, labour-surplus, resource-poor variety in which the vast majority of the population is typically engaged in agriculture amidst widespread disguised unemployment and high rates of population growth' (Ranis and Fei 1961: 533; also see Ghatak 1978; Lewis 1954). The characteristics of underdevelopment that China faced during the early reform period were reflected in a number of important factors.

First, there was the unfortunate combination of relatively limited natural resources, an already dismal state of the environment and a huge and far from stable population (see Smil 1993). Second, over 80 per cent of the population were rural with very low per capita GDP. As Yao states:

> By the end of [the] Cultural Revolution, three quarters of the rural population (570 million) lived in absolute poverty and did not have sufficient food to eat or warm clothing to wear. …[R]ural per capita disposable income in 1978 was only 285 yuan in 1990 prices. This was significantly lower than the official

poverty line of 318 yuan, and it was much lower than the poverty line of 454 yuan applied by the World Bank.

(Yao 2000: 448)

Third, China had a long history of rural underemployment (Buck 1937, Rawski 1979, Huang 1985). According to Rawski (1979: 124–5), a very important cause of rural underemployment under Mao's regime was industry's relatively slow absorption of labour. This in turn resulted from the then government's development strategy which emphasized heavy industry, and the command economy implemented a resource allocation system, which led enterprises to favour the use of capital rather than labour to expand output. Limited creation of employment outside agriculture forced more labour absorption into the farming sector (this latent pool of labour was later to play an important part in the expansion of TVE coalmines). Agriculture, it is estimated, had absorbed 97.3 million workers between 1957 and 1975, or about two-thirds of the overall labour force increase during these years (Rawski 1979: 125). Perhaps unsurprisingly, available data show that overall factor productivity dropped substantially between 1957 and 1975 in China's agricultural sector (Rawski 1979: 128). These findings proved that while there was already underemployment in agriculture, the labour absorbed in agriculture was not able to increase productivity but rather added to even more underemployment. The disguised unemployed rural labour force was estimated still to be as high as 100 million by the 1980s (Du 1989: 7; Chen 1989: 212).

After the Cultural Revolution there was a huge debate in China about how best to shape the challenges from such underdevelopment (Naughton 1995: 59–76). The debate ended in April 1979 with a fundamental reorientation of economic development strategy.

> The basic idea of reorientation was to reduce heavy industry investment and shift resources to agriculture and consumption, thereby also moving the economy onto a more moderate growth path with less strain on available resources.
>
> (Naughton 1995: 76)

Eventually, quite unlike the earlier Mao campaigns, the relatively modest and silent rural reform was the spark that set development alight. China's industrial reform did not really fully start until 1984, six years later than agricultural reform.

Some very important literature on the dual economy enables us to understand why this development strategy shift was so important for China's economic success. Lewis (1954, 1979) firstly pointed out that since the majority of the labour force in less developed countries (LDCs) consists of farm people, who also have the lowest incomes, the standard of living of the great bulk of the population can be raised 'only by raising farm income'. However, the closely related problem is how to raise farm income when there is a large amount of disguised unemployment crowding the rural market.[3]

Disguised unemployment means that, even with unchanged techniques of agriculture, a large part of the population engaged in the agricultural sector could be removed without reducing agricultural output (Nurkse 1953: 32). While Nurkse believed that development could be initiated and accelerated in these countries, by forming capital through the employment of redundant rural labour, and by permitting better organization through 'consolidation of scattered strips and plots of land' (Nurkse 1953: 33), Eckaus (1955) responded that many underdeveloped nations have less capital than is required.

Lewis (1954) made the greatest contribution to this area of research which was the foundation of all subsequent 'two-sector' models. His unlimited supply of labour dual-sector model presents a most important analysis of the relationship between the subsistence and capitalist sectors of the rural population in developing countries. In his model, surplus labour is available in both rural and urban sectors, from subsistence agriculture, casual labour, petty trade, domestic service, wives and daughters in the household, and the increase of population. If the country is over-populated relative to its natural resources, the marginal productivity of labour is negligible, zero, or even negative. Surplus labour is disguised in the sense that everyone is working, but if some proportion is withdrawn, output will not fall; the remaining workers will just work harder.

Why do both rural and urban workers not receive their marginal product but can receive a higher traditional wage? Lewis believed that the average product per worker in agriculture determines the traditional wage. In peasant agriculture, each family member receives the family's average product regardless of his or her contribution. Since there are no opportunities of receiving a wage higher than the average product on the family's farm, there is no motivation to leave the farm and the average product will be greater than the marginal product.

However, the inequality between rural and urban income will eventually lead to migration from rural to urban areas, which determines that labour employed in the capitalist sector will also be paid the traditional wage as long as there is a surplus of labour in the subsistence sector. The low and constant wages permit large profits for potential reinvestment in the capitalist sector. The economy grows at a faster rate, because profits grow relative to the size of the capitalist sector and in increasing proportion of national income in reinvestment. The reinvestment in the capitalist sector will take more people into capitalist employment out of the subsistence sector. The surplus is then larger still, capital formation is still greater, and so the process continues until the labour surplus disappears. Meanwhile, the government can assist this transformation in many ways.

The message of the Lewis model is very clear for developing countries. First, development commonly relies heavily on transforming the rural surplus labour force from farm to non-farm sectors, not only in the sense of improving 'the standard of living of the great bulk of the population', but also in the sense of the nation's final industrialization. Second, surplus labour can be used instead of capital in the creation of new industrial investment projects, or it can be channelled into nascent industries, which are labour intensive in their early stages. Third, the government, if it can, should play an important role in initiating and fostering such a transformation.

Ten years later after Lewis's research, Fei and Ranis (1963: 283) presented a similar strategy for underdeveloped economies: 'the heart of the development problem [in such an economy] lies in the gradual shifting of the economy's centre of gravity from the agricultural to industrial sector through labour reallocation'. However, with many other critics of the Lewis model, Fei and Ranis assume that Lewis did not pay enough attention to the importance of agriculture in promoting industrial growth. Thus, the Fei and Ranis model (1964) highlighted three major points. First, the growth of agriculture is as important as the growth of industry. Second, the growth of agriculture and industry should be balanced. Third, the rate of labour absorption must be higher than the rate of population growth to get out of the Malthusian nightmare.

Such criticism is not entirely accurate, as Lewis actually emphasized the necessity 'to make the countryside economically viable, with a larger cultivated area, with rising productivity on the farms, more rural industry, and better social amenities'. He even claims that 'it is agriculture which finances industrialization.'[4] Nevertheless, these criticisms should be a reminder for developing countries to pay attention to both agricultural and industrial prosperity and the transformation from rural to non-rural development because eventually industrialization is inevitable for further development.

These dual-economy models illustrate the importance of China's shift in development strategy from heavy industry to improving people's living standards, for which agriculture and rural development needed more attention. As already stated, the economic development of China took off after that of rural development. The household responsibility system (HRS) first caused a sharp increase in labour productivity, which in turn turned the rural labour force to more productive uses, making the disguised unemployment problem more explicit. Further, township- and village-owned enterprises (TVEs) were encouraged to grow. Since the labour force used by TVEs was 'disguised unemployed', so their withdrawal from agriculture did not cause agricultural output to decrease. By contrast, '[g]rain output increased from 305 to 407 million tons over the period 1978–84. Real per capita income more than doubled' (Yao 2000). Moreover, the TVE sector was allowed to develop and, by the early 1990s, became more important than the agricultural sector in terms of output value. By 1995 'about 29 per cent of the total rural labour force was employed in the rural non-agricultural enterprises, while 56 per cent of the value of national industrial output was from the rural industrial sector' (Meng 2000: 36). Finally, the industrial sector had also been developed by absorbing cheap labour and raw materials from agriculture and now by being forced to compete with TVEs.

Clearly, rural development, rural–urban transformation and TVE development were keys to improving economic development and people's living standards, and thus instrumental in the success of reforms in China.

However, the challenge from the force to promote development is still critical in the twenty-first century. First, the level of development in China is still very low. By 2000, the average per capita GNP was $840, below the $1,230 for low-medium income countries, and far below the $5,170 world average, ranking 141 in the world (World Bank 2002: 19–20); the net annual rural income was only $194 (SSB 2001); and 114 million people still lived below the poverty line (Yao 2000).

Second, by 1999 there was still a labour suplus of 120 million in rural areas and population growth would lead to a predicted additional 28 million between 2000 and 2004 (Shi and Zhao 1999: 50). It is predicted that rural underemployment will become more serious in the future as demand for labour will decrease. The capacity of TVEs to absorb the rural labour force may decline because they have seen limited development since the mid-1990s and there is no guarantee that their rapid growth will resume. The increasing capital intensity of, and the technological innovations adopted in, the agricultural sector will also lead to less demand for labour. There is also a trend for the urban industrialized sector to absorb less rural surplus labour because of its increased use of more advanced technologies. Moreover, these last two trends will very likely be strengthened due to the impact of China's joining the WTO.

Third, in the past, economic development in China, though impressive at times, has seemed unsustainable. China is currently meeting the daunting population challenge by producing 360 kilograms of grain per capita, but the extent to which it can do so sustainably into the future, and the extent to which the agricultural sector will promote broader development, will depend greatly on the success with which current land-tenure reforms are carried out (Prosterman 2001: 80). In addition, it depends on the effectiveness of dealing with those consistent 'old' problems such as land degeneration, deforestation and air pollution (see Smil 1993).

Finally, past economic development has also incurred new challenges for improving human, social and political development. Rapid income growth in China has been accompanied by rising inequalities. According to official estimates, China's Gini coefficient rose to 0.45 by the end of the 1990s (Ma and Wang 2001: 43). Before economic reforms China was a more egalitarian society. Twenty years later income inequality is on a par with the income inequalities of her neighbouring countries in east Asia and the Pacific region (Yao 2000: 5). The rural–urban income gap has been responsible for half of the increase in inequality since 1985 (World Bank 1997a: 2–3).

It is commonly agreed by China's policy-makers that these remaining challenges imply that the country must make a greater effort to improve its rural development:

> In early industrializing countries, during the development period when their per capita GDP was over US$1000, they usually implemented policies that completely supported industrialization by giving way or sacrificing agriculture first, then letting the established industrialization protect or develop agriculture later. However, China can't follow this path because of its huge population, food demand, employment and environment. …China should industrialize while also developing agriculture … and also actively develop non-rural industry.
>
> (Shi and Zhao 1999: 96–7)

Transition

As well as facing the challenge of increasing per capita incomes, China has also faced the disadvantage of carrying out a 'transition' from the former command economy, as in many other communist countries in eastern Europe a decade later. The challenge of transition is not only located in that the former command economy system must be transferred, but also in how it is to be transferred. Kornai divided the numerous strategies recommended for, or implemented in, transition economies into two basic categories representing gradualism and shock therapy strategies respectively:

> Strategy A … emphasized the creation of favourable conditions for bottom-up development of the private sector: encouraging the launch of new firms by eliminating barriers to entry, guaranteeing the security of private ownership, enforcing private contracts, and applying affirmative action – cautiously – for example, through tax and credit policies … In contrast, strategy B's emphasis was on the rapid elimination of state ownership. It called for privatization primarily through some form of giving – for example, voucher schemes. The goals were dispersed ownership – the equal distribution to all citizens of state assets – and the development of 'people's capitalism'.
>
> (Kornai 2000: 1–2)

By this definition, China's transition experience so far has followed the pattern of strategy A, while Russia's might be best described as following strategy B. As Kornai (2000: 2) recalls, however, in the early 1990s, 'the vast majority' favoured strategy B – the rapid elimination of the state sector. This 'shock therapy' strategy called for former socialist countries to have a complete and rapid 'personalization' of property rights and transfer of state-owned assets into the hands of 'real flesh and blood persons' (Kornai 1990: 50, 57, 70, 75, 85). According to this strategy most of the large-scale SOE sector would need to be closed down. Attempting to support and reconstruct the large-scale state sector was regarded as a waste of resources. Industrial growth was visualized as coming mainly from new small and medium-sized enterprises (Blanchard *et al.* 1991: 64–5).

Ten years into transition in eastern Europe and twenty years in China, even Kornai (2000: 2) confessed that 'experience has proved that strategy A was superior to strategy B'. Åslund (2002: 93) also admitted that 'the Chinese model stood out as a successful model of postcommunist economic transition'. However, this change of view arose primarily from the recognition of China's economic achievement *vis-à-vis* Russia and other transitional economies, rather than from understanding the basic necessity of such a strategy. Indeed, the path followed by China has confounded predictions of most earlier writers. The lack of understanding of the need for gradualism in China is best reflected by the consistent criticisms that China kept large loss-making SOEs functioning, which is usually taken as major evidence for the failure of her gradualism. Numerous scholars have been telling the story of 'inefficient institutions causing poor financial performance' (e.g. Cao *et al.* 1997; Fan and Woo 1996; for comments on this 'story' see Chang and Lo 2002).

In the view of the transition orthodoxy, China's SOEs have lagged far behind non-state enterprises. The absence of privatization on the grounds of preserving social stability may also have overlooked the social tensions created by this asset stripping, high levels of corruption and macroeconomic instability (caused by the ownership structure of SOEs) (Sachs and Woo 1997; Sachs *et al.* 2000). It has been warned that 'delays in reforming large enterprises and the financial sector [in China] have been costly, and constitute a major challenge for the future' (IMF 2000: 120).

Entering the twenty-first century, SOEs remain in financial difficulty. Meanwhile, plans to protect unprofitable SOEs might be incompatible with China's accession to the WTO. Dealing with the problems of SOEs therefore comes at a critical time. Calls for dramatic changes in SOEs, including the liberalization of the state-owned financial and insurance sector, naturally have become stronger (Wu 2001a).

This point of view, however, misses two important points. First, gradualism is an unavoidable choice for China despite the problems faced by SOEs. Second, SOEs' current difficulties prove not the failure but the success of gradualism. While the second point will be elucidated in Chapter 4, this following section briefly explains the first point.

The Chinese government under Deng's regime firmly initiated transition to promote economic development. Facing the severe situation of rural under-employment and poverty, the post-Mao government had to emphasize a gradualist transition – 'the creation of favourable conditions for bottom-up development of the private sector' and 'encouraging the launch of new firms by eliminating barriers to entry' (Kornai 2000). What China created, however, was not only the private sector, but also the TVE sector. This was why TVEs were encouraged to develop after the rural household system reform, which greatly promoted the rural development. There was no reason for China to deploy a 'shock therapy' transition model, as deployed in eastern Europe, which mainly focused on urban transition. While most eastern European countries were highly industrialized [5] with 75 per cent of their labour force employed in non-agricultural communes, over 80 per cent of China's population were rural (Hussain 1983; Sachs and Woo 1994).

That the transition was to promote economic development also determined that, immediately after the reform in 1978, there was no need to privatize SOEs while encouraging the growth of TVEs. This is because SOEs produced the bulk of national output at the beginning of the reforms, and were the only valuable source of productivity under government control. Meanwhile, TVEs not only provided more and more new productivity, but they also competed with SOEs and stimulated them to improve their productivity. The economy was therefore enormously improved by the combination of the productivity from these two groups – SOEs, both old and improved, and TVEs. Many other scholars have found that the new non-state sector generated growth and development without detrimental effects on the old state sector (Chen *et al.* 1992; Amsden *et al.* 1994; Goldman 1996; Nolan 1995).

Transition might hold great challenges for China, which also determined that gradualism would be superior to shock therapy strategy. Transition might, first of all, be a challenge for the government's finances because of its huge costs. Taking the costs of transforming SOEs as an example, they at least included payments for redundancies from SOEs, such as basic maintenance payments for the first three years and unemployment payments for the following two years; the cost of clearing the SOEs' bad debts; payments to local communities as a precondition for them to settle local unemployed, and many more. However, the government's financial capability to support these transition costs proved inadequate, especially during the early period of reform.

According to a most authoritative report from the Development Research Centre of the State Council of China (Ma and Wang 2001: 84–5), the government's inadequate financial capacity resulted from the consistent decline in the ratio of the government revenue to the GDP, and the ratio of the central government revenue to the total government revenue since the reform. While the first ratio started to increase since the tax reform of 1994, reaching 15 per cent in 2000, the second ratio did not increase until 1998, reaching 51 per cent in 2000. It was estimated that government revenue would decrease during the Tenth Five-year Plan to be 3–5 per cent lower than that in the Ninth Five-year Plan.

While government revenue has been declining, government expenditure has been increasing rapidly since the 1990s. More expenditure has been and will still be made on the following (Ma and Wang 2001: 190–1):

1   clearing bank bad debts, amounting to a few hundred million yuan from 1996 to 2000;
2   restructuring costs, such as subsidies to the Western Development projects;
3   social investment to stimulate aggregate demand, as non-official investment is still weak;
4   compensation or emergency payment due to social instability resulting from transition and redundancies.

It is not surprising that, as a result, government finances were in deficit in most years, which might be a fundamental cause of the government not being able to afford the huge costs of transition. Taking pension reform as an example, the costs of the current pension system for China, both today and in the future, are substantial. In 2000 the system had a deficit of at least 87 billion yuan ($10.5 billion), not including the 100 billion yuan ($12.1 billion) already 'borrowed' from individuals' pensions (Chen *et al.* 2001: 2).

Transition would also be a challenge for socialist legitimacy. Transition will lead to large-scale redundancies, bankruptcies and privatization. But China has followed a socialist ideology for decades while still implementing 'market socialism'. According to the well-known Chinese economist Wu Jinglian (2001b), this is equivalent to a 'market system plus social justice'. In this socialist legitimacy, SOE workers have been officially recognized as the 'leading class' and hence traditionally enjoyed guaranteed jobs, free accommodation and healthcare. The 'leading class' cannot be left homeless overnight if the government hopes to maintain its legitimacy. It therefore needs time to change people's ideology and attitudes before they will accept the transition.

Transition would be a challenge for institutional structures, as China's social security system, legal framework and financial system are all unable to supervise a dramatic transition such as comprehensive privatization. Taking social security reform as an example, as China is such a regionally diverse country any reform must allow policies to cover different types of workers in regions with different levels of economic growth and development. Consequently, social security arrangements have to vary across the country. In addition, any new social security system must be developed and implemented together with the improvement of its economic infrastructure, including employer and employee identification procedures, financial reporting, communications, computing, and accounting systems. Reforming the financial and legal system has similar difficulties. These are huge tasks and it is questionable whether the country will be able to achieve them quickly at all.

Transition would be a great challenge for social stability. Transition will result in reallocation of priorities, which may incur the resistance of those who lose; transition is not always positive for development, and frequently brings about reductions in productivity, as shown by the dramatic transitions under Mao's regime and the most recent one in the former USSR. However, since people's expectation of transition is to 'make things better', social instability usually occurs as a result of people's resentment of their losses and of any increases in inequality during the system change.

Taking the nation's development requirements into consideration and fearing these challenges from transition, China employed a gradualist transition. TVE development not only improved her economy, but created a competition for SOEs and forced them, in turn, to be transformed. Moreover, the improved economy reduced difficulties of further transition and thus created a beneficial chain effect. This is the biggest difference between Chinese and Russian transitions. China conducted transition accompanied by development and it was thus far less painful. Russia and eastern Europe conducted transition without the prospect of forward development momentum by adopting 'shock therapy'. This is the key reason why China developed while at the same time maintaining her stability, while Russia fell into economic stagnation and social instability. As Wang (1998: xx) notes, 'growth itself can facilitate and promote system changes'.

The current challenges from transition are still overwhelming in China. The economy has turned from one of shortage to one of surplus. SOEs' difficulties as a result have become more prominent, and more intense competition from global big business has also forced the government to reorganize SOEs as quickly as possible. However, once again, the dramatic transition for SOEs requires government funds to pay for transition costs, a social welfare net, delineation of clear property rights, and the independence of the financial sector. Unfortunately, all these conditions have still not been met. This indicates that, despite the well-known SOE problems, dramatic transition is still not an option.

Globalization

The change in China's foreign policy from the closed door to membership of the World Trade Organization (WTO) shows the importance to China of its integration into the world economy. China opened its doors by recognizing the miserable consequences of the closed-door policy of the 1950s, 1960s and 1970s. Before the reform China regarded trade as a necessary evil. Imports were used to achieve greater self-sufficiency and exports only served to pay for imports. Trade was limited as much as possible and was not seen as having merit in itself or as a viable strategy for her economic development. Her position in the international economy was negligible. Her share in world trade averaged only 0.6 per cent (Chai 1997: 139).

The open-door policy was an integral part of China's new development strategy, in an initial attempt to ease domestic competition for raw materials, provide employment for China's huge surplus labour, improve China's industrial competitiveness, and take advantage of perceived international opportunities (Yang 1991: 63; Chai 1997: 139). Between 1979 and 2002, China made use of $623 billion of foreign capital, of which $446 billion was foreign direct investment (FDI) (CSA 2003: 168), the largest in the world apart from the USA. There is an argument that China's economic success should be ascribed to its engagement in world trade, which has been taken as further evidence to encourage China and other developing countries to integrate completely with the world economy (e.g. Zhang, Zhang and Wan 1998).[6] Partly as a result of this, China finally joined the WTO in 2001.

However, there is no full recognition in China of the negative effects of such an engagement. While many people are expecting the promised benefits from the accessionn to the WTO, there is an urgent need for more people to realize fully that the challenge from globalization will be real and severe.

From Adam Smith (1776) to David Ricardo (1817), the significance of inter-national trade for economic growth has been recognized. They argued that in an open economy, development would proceed faster and more efficiently. However, for many other classical economists at that time and for the governments of European countries in the eighteenth century, openness did not mean free trade. 'They favoured mercantilist trade policies, and they believed that initial import substitution to protect infant industries, combined with selective export promotion, was needed to initiate development' (Adelman 2001: 113). In practice, Smith's theory was not popular for almost 100 years that following his death. Between 1750 and 1850, the British economy was highly protected and consequently achieved great prosperity. Only after British manufacture had become the world's most advanced, did Smith's theory become admired. Mobilized by free trade imposed through diplomatic and military pressure, Britain's products destroyed national industries all over the world. Impacted by the loss of her national industries, the USA was strongly opposed to the free trade theory, and highly protective of its industries from the late eighteenth century to the early twentieth century. After the USA developed, free trade theory was not only advocated, but also encouraged in other developing countries.

Neoclassical trade theorists (Krueger 1979, 1983; Bhagwati and Krueger 1973; Bhagwati 1985), however, despite this historical evidence, put considerable emphasis on free trade, which can basically be understood as international trade without any government intervention. They believe that the only thing that the government needs to do to attain an autonomous, sustained-growth path to remove barriers to international trade in commodities. The implication for the development policy in developing countries should be a free-trade regime, characterized by low or negligible impediments to imports, and incentives for export sales equal to the incentives for domestic market sales. By expanding the proportion of the economy directly subject to international competitive pressures, the government's own ability to impose 'political' prices is weakened and hence producers' uncertainty about government policy is reduced (Yang 1991). Free-trade theories have been pushed to extremes since the 1990s, especially after the emergence of the Washington Consensus (see Williamson 1993, 1997; Gore 2000).

In practice, it is believed to be essential for developing countries to move 'as rapidly as possible to a transparent and decentralized trade and exchange rate system, in order to hasten the integration … into the world economy' (IMF *et al.* 1990: 17). Integrating with the world economy is no longer a matter simply of dismantling barriers to trade and investment. Countries now must also comply with a long list of admission requirements, from new patent rules to more rigorous banking standards; from public sector reforms to humanitarian rights records. For developing countries short of investment funds, accepting some of these conditions is a prerequisite to receiving loans and investments from international organizations.[7]

These mainstream theories overlook the negative effects of free trade on development for poor countries. These have been exposed by among others Marx and Engels[8] (1987, 1888), Polanyi[9] (1957) to Ishikawa (1967). Export-led growth theories (e.g. Amsden 1989; Wade 1990; Chang 1996), especially, deploy the most recent development experiences of Japan, South Korea and Taiwan to demonstrate that the effect of international trade on economic growth is dynamic. International trade was important for the economic success of Japan, South Korea and Taiwan, but the governments deployed a series of industrial policies to effectively control the trade in terms of its time, speed, amount and orientation. According to Wade (1990), it was not the free trade, but the government's successful 'governing market' that was key to their economic success.

Particularly since the 1990s globalization, in general, (see Held *et al.* 1999) and a global big business revolution and financial liberalization, in particular, (see Nolan 2001a, b) have created many new challenges for developing countries. The revolution has brought a dramatic growth in the business capability of leading international firms, through concentrating on their core business, enhancing their brand, and massive spending on research and development, and information technology. The boundaries of large corporations especially of those multinational corporations (MNCs) have also become significantly blurred (for more detail see Chapter 5).

There are numerous challenges for developing countries from this massive revolution. It would be a challenge for uncompetitive industries and companies in developing countries owing to the increased size and power of these large corporations. It would be a challenge for the usually surplus labour force and immature social security system in developing countries, because the applied advanced technology, equipment and management will usually lead to redundancies or labour force mobility. Moreover, it is a particular challenge for the financial systems of developing countries, which may be affected or may even collapse due to the changes of international financial circulation. This has been seen in almost all the financial crises which have happened recently, including the latest in Argentina (Myint 2001: 526–7).

The Tenth Session of the United Nations Conference on Trade and Development held in Bangkok in 2000 provides comprehensive evidence for the above arguments. It argued that globalization is led by the developed economies, and bears the imprint of the strongest economic power in those developed countries, while the problems of developing countries, which are more urgent and compelling because of their relative poverty, have been largely overlooked. Globalization has brought increasing uncertainty to the global economic environment, but developing countries have little control over the process. Moreover, it is much harder for developing countries to catch up (UNCTAD 2000).[10]

The tragedy that occurred at the Ministerial Conference of the World Trade Organization held at Cancun in Mexico on 10–14 September 2003 once again provides horrific evidence of how free trade damages interests of less developed countries. One of the key issues of the Conference was the continuing negotiation on agriculture and services. This provoked demonstrations in Cancun by farmers from all of the world. On the second day of the conference a South Korean farmer, Mr Lee, even stabbed himself to death in order to prevent the unfair trade resulting from subsidies to farmers in developed countries. However, with no concessions being made by the developed countries, once again there was consequently no agreement reached at this conference.

Fortunately, China has not followed conventional theories of close integration with the world economy. China's strategy to use foreign loans cautiously but to encourage comparatively less risky FDI was clearly established since the very beginning of the reforms (see Bell *et al.* 1993).[11] Moreover, while other Asian developing countries liberalized financial systems and put foreign loans to unproductive uses, China better controlled the use of loans (it basically invested in infrastructure rather than real estate). This cautious policy helped China not only to overcome the difficulties of accessing the market for medium- and long-term funds after the 1989 Tiananmen Square students' movement event, but also to avoid suffering from the Asian financial crisis in 1997. Finally, China was aware of the importance of strengthening itself for the final face-to-face confrontation with the global giants after accession to WTO, and intended to build up a list of globally competitive indigenous large firms to fight global giants by deploying a series of industrial policies.

However, despite these previous successes, the world big business revolution will have particular challenges for China in the twenty-first century. Well before China joined the WTO, almost all Fortune 500 firms were active in China (SPC

1999a, b). By 2003, three-quarters of the top 500 USA companies have entered China. Unfortunately, China has a chronic problem of diseconomies of scale and many firms are too small to compete with global giants. Despite important progress, even after two decades of industrial policy, not one of China's leading enterprises has become a globally competitive corporation, with a global market, a global brand or cutting-edge technologies and product development capabilities.

Joining the WTO will make it even harder for China to establish large indigenous firms through industrial policy. The US–China Agreement signed in 1999 as a key agreement whereby China joined the WTO in 2001 is the most detailed agreement yet signed by any country on its entry to the WTO. The agreement in itself constitutes a massive programme of economic reform. Nine hundred Chinese laws will need to be changed and/or adapted for her to enter the WTO. Within the international environment, as shown before, globalization has made it difficult for developing countries to implement their domestic national policy (see Griffin 1996: 127–8).

Inter-relationship of the three challenges and the role of the state

Development is a paramount challenge and the ultimate goal in China. The process of both transition and integration with the world economy ultimately should help to serve this goal. This could be considered the most important principle in understanding the inter-relationship between the three challenges.

Development is a paramount force and ultimate goal in China because of the status of her underdevelopment as described above. Her low level of economic development, however, constrains the speed and degree of transition and limits the capacity to compete with companies on an international playing-field. It also limits the way in which the benefits from globalization (such as following high technology and acquiring new knowledge) can be reaped. Because China also needs to cope with challenges from transition and globalization, development becomes even more important.

Transition from the formerly unproductive command economy is necessary for China. But transition does not automatically promote development; sometimes it may reverse it, which is, unfortunately, the case in Russia. Development is supposed to make things better rather than worse (Clark 1999); the transition process should be taken cautiously with a precondition of not damaging too many prospects for economic growth.

Globalization in general and integration with the world in particular has assisted China's development. However, as shown in the last section, globalization and the big business revolution have placed developing countries, including China, in a very disadvantageous position. China faces fiercer competition from foreign companies on both the global playing-field and within China's domestic market. It is a complex question asking exactly what kind of effect this will have on China's development in the longer term, particularly on its efforts to build indigenous industrial groups.

In practice, China has been exerting herself to use the positive effects of transition and world integration to promote development which is basically what has determined her pursuit of all three goals together. While the available resource is limited, to meet one challenge may not make it possible to meet another, so it is therefore very difficult to deal with the competing interests of the three subjects. Therefore a powerful government is needed in China. It needs the government to design the focus of development at different periods; to control the content, speed, and size of transition to best serve the purpose of development, and to design industrial policy to improve national industry and to enhance global competitiveness. Moreover, it needs a strong government to cope with the conflicting interests of development, transition and integration with the world at the same time time, to meet one challenge but ignore another, or sacrifice one to achieve another.

Neoclassical economists believe that state interventions usually make things worse (see Krueger 1974). Therefore, the state should leave private producers operating through market mechanisms to supply all but certain public goods. This sort of view reached a peak after the emergence of the Washington Consensus (Williamson 1993), which widely believed that the communist bureaucracy could not lead a sufficient movement towards a market economy so as to cross a minimum 'threshold' level of market activity (Åslund 1991, 2002; Kornai 1990, 1992a, 1992b; 1994; Sachs 1992a, 1992b, 1992c; Sachs and Woo 1994, 1997; Sachs *et al.* 2000). In this view, transitional economies would automatically grow once they ousted their communist regimes, established completely free market systems and integrated closely with the world economy.

However, this sort of miracle did not occur in the countries where the state's role was reduced to a minimal level, and which accepted and implemented the orthodox recommendations (including the former USSR and many east European countries). In contrast, China's economic success was more precisely based on strict political controls. The weaknesses of the conventional theories, as Stiglitz (1999a) argues, stem from a 'misunderstanding of the foundations of a market economy as well as a misunderstanding of the basics of an institutional reform process'.

First, there is no doubt of the merits that the market mechanism possesses. However, 'market failure' is widely admitted and state intervention has been found necessary and effective in cases of market failure, particularly regarding the problems resulting from public goods, a non-competitive market and externalities.

Second, whether, or to what extent, the merits of a market can contribute to an economic system depends on many other conditions of the economy. In many developing countries these conditions may remain blank, and many assumptions of neoclassical market theories simply do not exist,[12] in which case the merits of the market system may be less. This is why Killick (1989) believes that neo-classical market theories have a value bias on less developed countries. Steinfeld (1998) and Stiglitz (1994, 1999b, 2000) have similar arguments.

Third, in contrast to mainstream theories, developmental state theories have found government intervention and industrial policy were fundamental sources for the economic miracle experienced by east Asian countries from the 1970s to 1997 (Amsden 1989; Wade 1990; Chang 1996, 2002). Although there have been found to be weaknesses and problems with state intervention, a wise approach is not to disregard the possibility of certain types of state intervention, but to combine the merits of both state and market. Industrial policy is just such a sort of state intervention and it has been successfully applied in both capitalist countries and newly industrialized economies (NIEs) (see Chandler 1990; Wade 1990; Chang 1996).

Fourth, not only may state intervention be necessary, much research even argues a strong state is actually needed to control internally the competing interests that arise in transition economies (see Nolan 1993, 1994, 1995, 2001a; Steinfeld 1998; Wood 1994). Transition will result in a redistribution of benefits and interests, with winners and losers, so there is a need for a strong state, therefore, to mediate between these parties and to compensate certain groups when necessary for the promotion of the transition process. Reforms must be forced through against strong opposition at times and this may only be possible by firm government.

Furthering the above argument, it is the belief in this research that the most important role for a transitional government today is to deal with the three stated challenges in a balanced, cautious and wise way. In contrast to Gorbachev's *perestroika*, which led to the dissolution of the USSR by liberalizing people's thoughts but without providing satisfactory economic rewards, Deng Xiaoping chose the 'Four Basic Principles'[13] to strengthen political control, while implementing economic reforms so as to enable people to enjoy economic rewards. The importance for this political control does not reside in whether the contents of the Four Principles are right or wrong, but in the fact that this is the most powerful way to prevent the nation becoming slack and people being distracted from promoting economic growth.[14] During the last two decades, China has been confronting the three great challenges quite successfully and the political control from the strong state was an important element in this success.

Conclusion

The challenges of *development, transition* and *globalization* are in conflict with one another and call for firm and effective government intervention so as to exploit their individual benefits, minimize the constraints and thus limit the consequences of this conflict.

2 The challenges facing China's coal industry

Coal is the food of industry.

Vladimir Lenin

Introduction

The Chinese coal industry might be the best showcase to demonstrate the three challenges China has been facing. The challenge from development has forced the coal industry to make tremendous efforts to fuel the nation's economy; to allow the problematic TVE coalmines to grow to assist the rural populations' desire to eradicate poverty; and to reduce its pollution to pursue sustainable development. The challenge from transition has been centred on the needs to transform the old command economy system and large loss-making SOE coal bureaux, and on the huge difficulties of planning this. The challenge from the trend towards economic globalization is reflected in the wider gap between China's coal companies and their counterparts among the leading global mining giants, the urgent need for China to build large globally competitive firms to pursue internationally competitive capacity, as well as the enormous difficulty to achieve this.

The following three sections will explore each of the three challenges reflected in the coal industry. A brief summary of the inter-relationship of the three challenges and an explanation of the role of the state in dealing with such inter-relationship follows. The information in this Chapter is based on an overview of the entire history of the coal industry under the PRC's regime from 1949 to the present, aiming not only to illustrate points made in Chapter 1, but also to provide background and guidance for understanding Chapters 3–5.

Development

The Chinese coal industry has been consistently facing the challenge posed by the country's development. This is reflected in three aspects:

1   As an irreplaceable energy source, coal holds specific obligation to fuel China's economy, from the period of severe shortage before 1990s, to meeting the

demands of the present day high GDP growth, and to ensuring sustainable development in the future.

2   As a major channel to develop non-agricultural business and increase income in most coal-rich areas, coal plays a significant role in absorbing the surplus labour force, in improving living standards, and in promoting local development.

3   However, as a major source of environment pollution impeding China's – and even the world's – sustainable environment, the use of coal has to be constrained and improved.

Fuelling and forging the economy

When Chairman Mao stated that the China when it was taken over by the Communist Party in 1949 was 'blank and poor', the new regime desperately wanted to find a way to improve the situation. Industrialization and catching up lost ground became priorities at the time. Coal was the key in realizing this ambition, basically because the nation's huge energy demand had to be met mainly by coal as there is no other a viable alternative, and coal is both an input and fuel for steel production. Frequent shortages also reinforced the significance of coal.

Coal has to meet China's huge energy demand and there are few alternatives. The country is relatively rich in coal, but poor in oil and gas. Coal accounted for 95 per cent of the primary energy consumption from 1952 to 1960; 80 per cent from 1961–70; about 70 per cent from the 1970s to the 1990s mainly due to the increased oil use after the discovery of the Daqing oilfield, and it still accounted for 69 per cent by 2002 (Table 2.1). Coal as a prominent fuel source generates around 80 per cent of the country's electricity at all times (Figure 2.1). The ratio of coal used in electricity generation to total coal output had increased from 17.98 per cent in 1980 to 60.85 in 2000 and 58.20 in 2001 (CEPY 2002: 642). The positive relationship between electricity generation and coal used in power generation is shown in Figure 2.2.

Coal is not only fuelling but also forging the economy, as it is also an important input of steel production and many chemical productions. Coal, because of its close relationship with steel production, was especially important for the Soviet Model which was adopted by the PRC in 1949 as the means of fulfilling its industrialization ambitions. The Soviet Model

> regarded the iron and steel industry as the basis upon which all economic development depended and the targets set for them virtually determined the targets set for all the other industrial sectors. Necessarily, planning for the coal and iron and steel industries was carried out simultaneously.
>
> (Thomson 2003: 37)

As the PRC took over only 320 coalmines with an output of 32 Mt in 1949 (CCEC 1999: 223), the construction of coalmines and recovery was emphasized. Of 156 projects for which the Soviet Union and the other Eastern Bloc countries

Table 2.1 China's energy production and consumption structure 1978–2002

Year	Energy output (Mtoc)	Structure (total output =100)				Energy consump-tion (Mtoc)	Structure (total output = 100)			
		Coal	Oil	Gas	Hydro-electricity		Coal	Oil	Gas	Hydro-electricity
1949	23.74	96.3	0.7	0.0	2.9	–	–	–	–	–
1953	51.92	96.3	1.7	0.0	2.0	54.11	94.3	3.8	0.0	1.8
1959	271.61	97.0	1.9	0.1	0.8	239.26	94.7	4.1	0.1	1.1
1969	231.04	82.2	13.1	1.1	3.2	227.30	81.9	13.8	0.8	3.5
1978	627.70	70.3	23.7	2.9	3.1	571.44	70.7	22.7	3.2	3.4
1980	637.35	69.4	23.8	3.0	3.8	602.75	72.2	20.7	3.1	4.0
1985	855.46	72.8	20.9	2.0	4.3	766.82	75.8	17.1	2.2	4.9
1986	881.24	72.4	21.2	2.1	4.3	808.50	75.8	17.2	2.3	4.7
1987	912.66	72.6	21.0	2.0	4.4	866.32	76.2	17.0	2.1	4.7
1988	958.01	73.1	20.4	2.0	4.5	929.97	76.2	17.0	2.1	4.7
1989	1016.39	74.1	19.3	2.0	4.6	969.34	76.0	17.1	2.0	4.9
1990	1039.22	74.2	19.0	2.0	4.8	987.03	76.2	16.6	2.1	5.1
1991	1048.44	74.1	19.2	2.0	4.7	1037.83	76.1	17.1	2.0	4.8
1992	1072.56	74.3	18.9	2.0	4.8	1091.70	75.7	17.5	1.9	4.9
1993	1110.59	74.0	18.7	2.0	5.3	1159.93	74.7	18.2	1.9	5.2
1994	1187.29	74.6	17.6	1.9	5.9	1227.37	75.0	17.4	1.9	5.7
1995	1290.34	75.3	16.6	1.9	6.2	1311.76	74.6	17.5	1.8	6.1
1996	1326.16	75.2	17.0	2.0	5.8	1389.48	74.7	18.0	1.8	5.5
1997	1324.10	74.1	17.3	2.1	6.5	1381.73	71.5	20.4	1.7	6.2
1998	1242.50	71.9	18.5	2.5	7.1	1322.14	69.6	21.5	2.2	6.7
1999	1100.00	68.2	20.9	3.1	7.8	1301.19	68.0	23.2	2.2	6.6
2000	1069.88	66.6	21.8	3.4	8.2	1302.97	66.1	24.6	2.5	6.8
2001	1209.00	68.6	19.4	3.3	8.7	1349.00	65.3	24.3	2.7	7.7
2002	1390.00	70.7	17.2	3.2	8.9	1480.00	66.1	23.4	2.7	7.8

Sources: CSA 2003: 136. SSB 1989: 81, 149.

Note

Mtoc = million tons of coal equivalent, '–' indicates data not available.

were providing assistance between 1950 and 1952, 25 were coal projects, of which 17 were completed and became operational (CCEC 1999: 136–7).

Experience shows that there is a certain ratio between coal consumption and steel consumption during the process of industrialization. In 1950 the ratio of coal consumption to steel consumption in Russia and the USA was about 16:1, but was 21.5:1 in China. This implied that the coal industry faced more pressure to match the speed of development of steel in China. Moreover, the recovery of steel production after the birth of the PRC was much faster than that of coal. During the time between 1949 and 1952 steel output increased at an annual rate of 104.4 per cent, while the annual increase rate for coal was only 27 per cent. The government had to plan in detail and tightly control coal supply to important industries and residential customers (CCEC 1999: 467).

The pressure became even more intense during the First Five-year Plan (1953–7) when the principal policy of the government was 'to take steel as priority' (*yi gang wei gang*),[1] during the Second Five-year Plan (1958–62) which introduced 'the campaign of all the people to refine steel and iron',[2] and the Third and Fourth

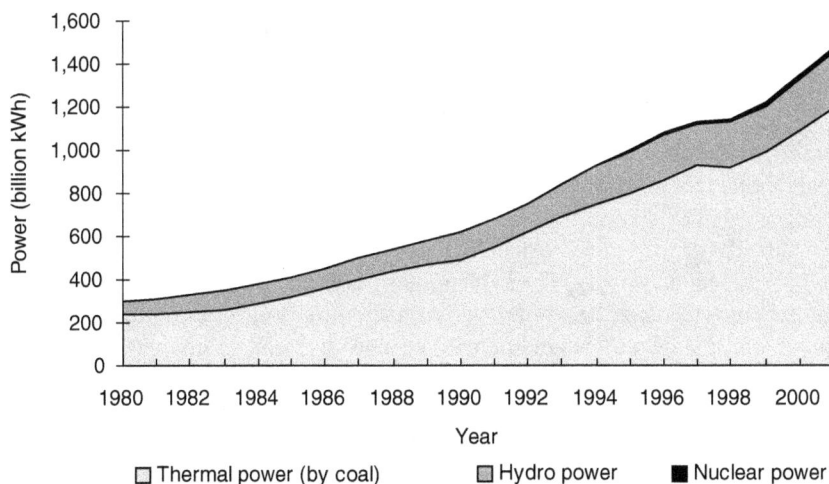

Figure 2.1 Electricity generation from different sources in China 1980–2001

Sources: SPC 1997: 136–9; CEPY 1998:10; 2002: 643.

Note
Statistics for nuclear power date from 1995 and a figure of 1.4 billion kwh has been more or less constant every year since then.

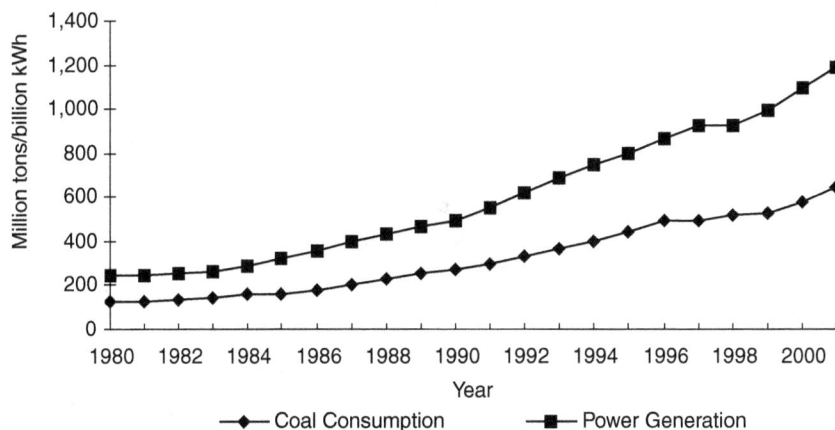

Figure 2.2 Power generation and its coal consumption in China 1980–2001

Sources: SPC 1997: 136–9; CEPY 1998:10; 2002: 643.

Five-year Plans (1966–74), which was also the time of the Cold War, when the steel and consequently the coal industry was ordered 'to prepare for the war, the famine, and to serve the people'.[3] After the economic reform, steel output increased dramatically from 65.35 Mt in 1990 to 181.55 Mt in 2002 (Yin 2003:4). Since the mid-1990s steel production alone consumes over 100 Mt coal annually, or one-tenth of the total (CEDREC 2001: 134). At present, when coal can generally meet the demand of the economy, a shortage of coke and coking coal is still the cause of fluctuation in steel output.

That the force of development presented a major challenge to the coal industry in China was typically demonstrated by the consistently severe shortages over decades and the consequent incalculable losses. It is no exaggeration to say that China's coal industry has a very troubled history. It seemed that the coal industry, during most of the period from 1949 to the present day, was like a dog chasing its own tail – the demands of an expanding economy and growing population chasing the never-adequate supply. The industry has never had the time to catch up with that demand, to repair and renew its infrastructure, to update its methods and equipment, to adjust to expanding markets and generally to improve itself. The vicious circle was one of ambitious economic development chasing ever-higher coal outputs, and this was followed by a consequent imbalance between extraction and coalface preparation, which was, in turn, followed by unsophisticated new mines being put into production, whilst at the same time economies imposed to meet ever-increasing demand led to skimping on investment in general and safety in particular. At first production increased, but rapidly declined later as the result of inadequate coalface preparation and the problems of trying to regain a balance between demand and production in order to meet ever higher targets for economic growth. And so the circle turned, supply never adequately meeting demand in an industry that needed time to reorganize itself from top to bottom.

Since 1957, when economic development started on a large scale, the demand for coal increased considerably. The Second Five-year Plan (1958–1962) had previously set the production target for 1962 at 190–210 Mt, and this already exceeded capacity. However in 1958 – the first year of the 'Great Leap Forward' – as the coal industry was ordered to exceed UK production in five years and to catch up with that of the USA within fifteen years, the original plan was revised, setting a target of 200 Mt by 1962, 500 Mt by 1967, and 800 Mt by 1972. This was revised again with targets of 300 Mt by 1962, 600 Mt by 1967 and 900 Mt by 1972. To meet these targets, numerous small and unsophisticated mines were opened up, causing many of the problems endemic to this type of mine. This sort of revision did not end until those imposed between 1963 and 1965.

Again, after the Cultural Revolution, as all other industries were recovered, demand for coal rose dramatically. In response, seven long-term plans were proposed for the industry between December 1976 and August 1978. The production target was set at between 580–700 Mt by 1980. In 1982 the Twelfth CCP Congress set itself the goal of increasing its GDP fourfold by 2000. Meanwhile, since the reform, the economy had shifted into top gear with an average annual growth rate of nine per cent. To keep up with this the Ministry of Coal Industry called for coal to double its output again to guarantee the quadrupling of the GDP by 2000. With this in view, coalfields in western China were planned in 1984, which included the setting up of the Huaneng Clean Coal Corporation, the predecessor of the Shenhua Group, whose responsibility was to develop the Shenfu Dongshen coalfield. China's first joint-venture coalmine, Antaibao, was established one year later in 1985 (CCEC 1999: 464).

Despite such pressure on the coal industry, no matter how much the priority and how much the effort, coal shortages were almost continuous until the mid-1990s. Even before the First Five-year Plan launched in 1955, coal production lagged far behind demand. Although the Ministry of Fuel Industry, then in charge of the coal industry, recommended an increase in production and intensification of coalface preparation, this preparation work was delayed in order to meet more immediate production needs. By 1957, due to the full operation of economic development, coal shortage was so severe that the government

> ordered all government offices, troops, and schools to reduce coal used for heating and cooking by 15–25 per cent and miners were asked to work overtime. ... While the 1957 budget [for coal] was 20 per cent less than in 1956, the coal production target remained unchanged.
>
> (Thomson 2003: 37–8)

The high economic growth rate since economic reform aggravated coal shortages. Two particularly serious shortage crises occurred in 1979 and 1988 (Ministry of Energy 1991, 1992). Taking 1988 as an example, the coal shortage was shown in four ways (Zhou *et al.* 1991:520):

1   power plants stopped generation because of coal shortages, and industries stopped because of electricity blackouts; and many factories worked for four days then stopped for three or stopped five days then worked for two;
2   the amount of coal in stockpiles rapidly dropped to the lowest point since 1983;
3   price rose dramatically, averaging 40 per cent increases at the end than the beginning of 1988, and up 130–200 per cent for out-of-plan coal in south-east China provinces;
4   even demand from key industries and enterprises was insecure.

The coal shortage, and the closely related electricity shortage, had a serious impact and coal was crucial to almost everything during the period of shortage. As Thomson records:

> Energy shortages have greatly hindered the industrial, agricultural, and social development of China. They have caused tremendous financial losses in foregone potential production and foreign investment. Capital spent on machinery that has been damaged by power failures, or which was forced to operate at only a fraction of its capacity, has been wasted. The lack of alternative to coal has perpetuated the gathering of burnable vegetation in the rural areas, resulting in the permanent loss of millions of acres of once fertile land through erosion, and thereby greatly contributing to recurring and devastating floods.
>
> (Thomson 2003: 1)

While the coal shortage frequently reinforces the picture of coal's significant role in China's development, the coal industry has been also stressed by the force of the development. This, in particular, necessitated the dramatic development of TVE coalmines. As shown later in Chapter 3, after the mid-1980s TVEs were encouraged to develop as quickly as possible. By 1996 TVE coalmines produced almost half of China's total output, from 20 per cent in 1981 (Table 2.2). This finally ended the history of decades of coal shortages.

Although the coal industry since the mid-1990s was, for the first time ever, able to meet demand (and partly because of this the government even started to close TVE coalmines), the challenge facing the coal industry from maintaining China's high speed of development is still hard. The growth of the economy and of the population would be the two fundamental factors to determine the challenge facing China's long-term development goal. China aims to become a country with

Table 2.2 Coal production in China, 1979–2002 (Mt)

Year	Total national output	Key SOEs		Local SOEs		TVEs	
		Output	% of total	Output	% of total	Output	% of total
1978	617.86	341.84	55.33	180.70	29.25	95.32	15.43
1979	635.54	357.77	56.29	171.46	26.98	106.30	16.73
1980	620.40	344.39	55.51	162.39	26.18	113.60	18.31
1981	621.63	335.05	53.90	159.99	25.74	126.59	20.36
1982	666.32	349.90	52.51	170.35	25.57	146.07	21.92
1983	714.53	363.12	50.82	181.34	25.38	170.07	23.80
1984	789.23	394.70	50.01	177.65	22.51	216.88	27.48
1985	872.28	406.26	46.57	182.78	20.95	283.20	32.47
1986	894.04	413.92	46.30	181.38	20.29	298.70	33.41
1987	928.09	420.20	45.28	181.13	19.52	324.70	34.98
1988	979.87	434.45	44.34	193.89	19.79	344.80	35.19
1989	1054.15	458.30	43.48	205.46	19.49	380.30	36.08
1990	1079.88	480.22	44.47	205.09	18.99	389.70	36.09
1991	1087.40	480.60	44.20	203.56	18.72	399.00	36.69
1992	1114.55	482.54	43.29	202.82	18.20	425.50	38.18
1993	1178.37	485.03	41.16	204.03	17.31	484.50	41.11
1994	1229.53	468.67	38.12	205.96	16.75	548.20	44.59
1995	1360.73	482.28	35.44	213.35	15.68	665.10	48.88
1996	1397.00	537.25	38.46	222.05	15.89	637.70	45.65
1997	1325.07	529.16	39.93	225.67	17.03	570.20	43.03
1998	1232.51	503.49	40.85	212.85	17.27	516.20	41.88
1999	1044.82	512.71	49.07	211.91	20.28	320.20	30.65
2000	951.06	528.15	55.53	191.28	20.11	201.63	21.20
2001	1105.49	618.48	56.00	223.16	20.00	263.85	24.00
2002	1393.35	711.63	51.00	263.45	19.00	418.27	30.00

Sources: 1) SPC 1997: 131; 2) CCIY 1999; 3) SCA 2000: 2; 2001: 1, 2003; 4) Ye and Zhang 1998: 401–2.

Note
The total of percentages for key SOEs, local SOEs and TVEs may not equal 100 because of the tiny ratio of other categories of coal producers.

medium levels of overall development by 2050. To achieve this goal, the economic annual growth rate has to be maintained at a certain rate. It is widely estimated that the annual GDP growth during 2001and 2010 will be around 7 per cent (e.g. Ma and Wang 2001: 9). Meanwhile, the population stood at 1.2 billion by 2000, which may reach 1.4 billion by 2010, 1.5 billion by 2020 and 1.45–1.58 billion by 2050 (Zhou and Zhou 1999). Currently, China's average per capita energy consumption is one of the lowest in the world (Table 2.3). It was 0.9 metric tons of oil equivalent in 1997, compared with 1.6 of the world average, 3.5 of the Europe and 8 of the North America (IEA, quoted from World Resource Institute 2003). Moreover, there are still 27 million rural inhabitants who have no access to electricity (CEDREC 2001: 9). There is a huge potential to increase the per capita energy consumption level if China achieves its aim of having a medium level of development.

Based on these two fundamental factors, and also taking into account changes in energy intensity and economic structure, it is estimated that China's energy demand will be around 2,700 Mtce by 2020 and around 4,100 Mtce by 2050[4], which might make China the largest energy consuming country in the world by that time (Horii & Gu 2001: 215).

Naturally, such high demand for energy will put most pressure onto coal. Over the years energy production structure has not changed so that China's capability to substitute other sources of energy for coal is very limited. 'China is an energy-scarce economy, with per capita energy endowments far below the world average' (World Bank 1997b: 48). China's per capita reserves of crude oil are just 3 tons, compared to a global average of 28 tons. Its per capita reserves of natural gas are just 1,416 cubic metres, compared with a global average of 28,400 cubic metres. The per capita reserves of hydropower are 1,603 kwh per year, compared with a global average of 2,909 kwh. By comparison, coal reserves are large and amount to 95 tons per capita, although still far less than the global average of 209 tons. Oil

Table 2.3 Energy production and consumption per capita (kg)

Year	Energy production per capita				Energy consumption per capita			
	Total	Raw coal	Crude oil	Electricity	Total	Coal	Oil	Electricity
	(kgce)	(kg)	(kg)	(kWh)	(kgce)	(kg)	(kg)	(kWh)
1965	263.2	324.4	15.8	94.5	264.3	320.0	19.0	94.5
1970	378.7	432.6	37.5	141.6	357.9	405.5	36.8	141.6
1975	532.0	526.0	84.1	213.7	495.7	498.8	73.1	213.7
1980	649.5	631.6	108.0	306.4	614.4	621.8	89.2	306.4
1985	818.9	834.8	119.6	393.2	734.1	781.2	87.8	394.2
1990	915.5	951.4	121.8	547.2	869.5	929.6	101.2	548.8
1991	905.2	675.1	122.5	588.8	901.8	959.6	107.6	591.3
1992	915.4	684.1	122.0	647.2	737.1	979.3	114.6	651.5
1993	937.1	697.4	123.2	712.3	984.3	1,021.7	124.9	715.1
1994	990.6	743.2	122.6	778.6	1029.8	1,078.4	125.5	777.0
1995	1,065.3	806.4	124.5	835.8	1088.7	1,142.7	133.3	831.9
1996	1,089.2	819.1	129.2	888.1	1141.2	1,188.7	143.2	884.1

Source: SSB 1996: 7.

reserves still remain uncertain. Currently China produces about 160 Mt of oil and imports 70 Mt each year, but it is estimated that by 2005 China will have a 100 Mt oil shortage when 38 per cent of oil demand will be met by imports. This degree of reliance on imports will be the third highest in the world, just behind the USA and Japan (CSSA 2000: 131). Interestingly, coal liquefaction has been planned as one option to solve the oil shortage problem and guarantee energy security in China.

Coal therefore looks like being the only reliable source of energy which, according to the updated overall research on a national scale, has proven reserves of 724 billion tons and predictable reserves of 5,570 billion tons by the end of 1999 (CCC 2000: 18). In view of China's current economic structure and financial capacity,[5] coal is widely estimated to account for no less than 60 per cent of her primary energy consumption by 2010 and 50 per cent by 2050 (e.g. Zhou and Zhou 1999: 15). Based on these estimates coal demand is predicted to be 1,200 Mtoc by 2005. To meet this demand, a capacity of around 100 Mt of new capacity needs to be achieved by establishing new coalmines between 2001 and 2005.[6]

Fighting rural poverty and underemployment

The coal industry in China, in terms of the development of its TVE mines with their unlimited supply of labour (see Chapter 1), might well be the best showcase in which to demonstrate the validity of the 'Lewis Model' of economic development.

Over 80 per cent of China's coal is located in western and north-western areas, which are historically poor and currently still lag far behind the national level of development. Frequently coal is the major source for local economic development and therefore becomes the only means of eradicating poverty and absorbing surplus labour.

Table 2.4 lists the ten provinces and autonomous regions which have the largest coal output and also the largest TVE coal output in their respective peak years. Their total coal output accounted for 73 per cent of the national total in 1997. However, their per capita incomes for the years 1978 (when the reform started), 1995 (when TVE output reached its peak), and 2002 (when the data was most recently updated) were almost all below the national average, especially for those located in the western regions. Only Liaoning, Shandong and Heilongjiang located in east or north-east China were better off.

As demonstrated below in Chapter 3 in the case study areas in Inner Mongolia, Shaanxi, Shanxi, and Heilongjiang, a strong drive to eradicate poverty facilitated the development of TVE coalmines. These coal-rich areas were not only poor, but also had huge surplus labour forces.

Before 1978, apart from local state-run coalmines, the central government only allowed communes and brigades to run collective coalmines, and did not allow the public to invest collectively in coalmines, let alone allow individuals to invest privately in mines (CCCIEC 1989: 105). However, poverty forced people to take risks. Well before 1978 individuals started to run coalmines, but usually with a

Table 2.4 Major coal and TVE coal areas and their per capita income

Province/ autonomous region	Coal output (Mt) 1997	TVE coal output (Mt) 1995	Per capita net income		
			1978	1995	2002
National level	1,325.25	659.00	133.57	1,220.98	2,475.6
Shanxi	330.38 (1)	157.29 (1)	101.61	884.20	2,149.8
Henan	100.28 (2)	43.40 (4)	101.40	909.81	2,215.7
Shandong	90.94 (3)	11.75	101.20	1,319.73	2,947.7
Inner Mongolia	79.08 (4)	30.18 (6)	100.30	969.91	2,086.0
Heilongjiang	75.47 (5)	25.10 (7)	167.90	1,393.58	2,405.2
Hebei	67.86 (6)	23.83 (8)	91.50	1,107.25	2,685.2
Guizhou	65.97 (7)	44.84 (3)	108.00	786.84	1,489.9
Sichuan	62.22 (8)	55.67 (2)	116.70	946.33	2,107.0
Liaoning	58.41 (9)	13.18	165.20	1,423.45	2,751.3
Shaanxi	49.58 (10)	19.75 (9)	133.00	804.84	1,596.3

Sources: CSA 2003:110; SSB 1999a:339; CCIY 1999: 21; Ye and Zhang 1998:59.

Notes:
1 National ranking of regional production is shown in brackets in column 2; and the TVE coal output ranking is shown in brackets in column 3.
2 Hunan ranked fifth for TVE coal output in 1995. Liaoning TVE figures are for 1997, and were well below 20 Mt so that it was not ranked. Shandong's TVE coal output was also not ranked.

'collective' title (Ye and Zhang 1998: 32). In spite of this, a private coal miner in Shanxi was punished by having both hands tied and being forced to wear a dunce's cap and walk around locally asking for forgiveness. It was said that he committed the crime of being a 'capitalist roadster'.

The policy was finally relaxed in the 1980s. Being aware of the contribution made by local coalmines in absorbing surplus rural labour, in producing much needed coal, and in eradicating local poverty, the government in 1981[7] first agreed to strengthen local state-run mines by subsidizing their losses. Second, in 1983 and 1984 individual and private coalmines were gradually allowed to develop so as to create a situation in which 'the state, collectives and individuals all work together to develop large, medium, and small mines'. Third, after 1983 local coalmines were allowed to transport and sell their coal outside their local areas, greatly benefiting both the areas of production and the areas of sale. Fourth, TVE coal and the coal outside the 'unitary distribution' in state-run local mines could be sold at market prices. Fifth, collectives and individuals were encouraged to use their own funds to build the infrastructure for coal production. From 1979 to 1984, local funds supplemented state funds so that 100 dedicated railways totalling 806 km of track and a group of railheads were built to transport coal from local state and collective coalmines. Meanwhile, many peasants bought trucks and other equipment to transport TVE coal. Finally, the administrative system in local state and TVE coalmines was also reformed. A responsibility contract system was applied to overcome equality and other problems.

TVE coalmines were encouraged throughout China as a result of these measures, It is difficult to imagine how strongly enthusiastic rural people were to improve their lifestyle and economy. Once the restrictive policy was relaxed, in 1983, 1984 and 1985 TVE coalmines output increased at an annual rate of 27, 21.58 and 23.43 per cent. By 1985 the number of TVE coalmines reached 63,000 (from 16,000 in 1982). Of the 63,000 TVE coalmines in 1985, 51,000 were collective and 12,000 private (Ye and Zhang 1998: 35). As described in detail in Chapter 3, TVE coalmines have greatly promoted the local living standards and economic development, and contributed to the ending of the country's history of coal shortage.

However, since 1998 tens of thousands of TVE coalmines have been forced to close. Does this mean that the challenge of rural poverty has been overcome and the Lewis Model is no longer valid? The answer is rather pessimistic. China's labour force accounts for 26 per cent of the world total. By the end of 1999, the total rural and urban labour force was 712 million, up 5.6 million on 1998. There are still large numbers of unemployed in both rural and urban areas.

Internal migration (*liudong renkou*) in China has been, and still is, a feature of the changes accompanying economic reforms. The number of migrants from rural to urban areas is enormous. According to official data from *China Statistical Digest 2000*, there were 101 million migrants from the countryside in 1999, up 5.6 million in 1998 when it was 9.5 million. According to the State Statistical Bureau (SSB), at the end of 1998, 60 million people from the countryside were living in urban areas (SSB, 2000). Unofficial data usually give a higher estimate. It is estimated that at any one time there are about 120 million permanent and temporary migrants in China (Croll and Huang, 1996). It is also estimated that the next few years will be the peak period for the rural labour force, with an increase of 8.57 million each year. This means that no fewer than 6 million members of the rural surplus labour force will become migrants (CASS 2001).

While rural migration pours into urban areas, an official report (*Outlook* 2003) showed that registered unemployment in urban areas already increased from 5.2 million in 1995 to 7.95 million in 2003. The urban registered unemployment rate is 4.2 per cent by June 2003. Taking 3.5 million laid-off SOE workers (*xiagang gongren*) and graduates into account, the unemployment rate in urban areas is estimated to be 7 per cent. It is also estimated that during the Tenth Five-year Plan 20 million people will enter the labour market each year and about 15 million will not be able to find jobs. Due to the large population base, China's unemployment is not only a structural one but also an absolute one.

More importantly, after more than two decades of development, the gap between urban and rural and between west and east have increased. This implies a great potential danger for social stability. As shown in Chapter 3, it is exactly the challenges of development – a stagnant national economy due to coal shortage, massive rural poverty and a large surplus rural labour force – that motivated the development of TVE mines. Meanwhile, case studies have found that TVE mines had become the most important channel in absorbing the surplus rural labour force and improving the local economy and living standards, and have, therefore, become the best option to close the gap between urban and rural areas and ease any social instability.

From this point of view, the present struggle between the closure or not of TVE mines actually reflects the historical challenge of development. After two decades of rapid development of TVEs through which many poor people were just beginning to have noticeably improved living standards, it needs to be asked whether the government should send these people back into poverty.

Ensuring sustainable development

While China's economic development is heavily reliant on coal and while the management of TVE coalmines is keen to expand in order to improve incomes and living standards for rural communities, China is required to produce and consume less and less coal because of concern for the environment and to ensure sustainable development, both for China and for the rest of the world.

Concern for the environment and sustainable development may greatly reduce the direct use of coal. Burning coal on open fires and in old-fashioned power stations was responsible for large amounts of atmospheric pollution through the generation of airborne particulates, for acid rain through the emission of sulphur dioxide (SO_2) and oxides of nitrogen (NO_x), and for global warming through the emission of carbon dioxide (CO_2) – the greenhouse gases. China has been accused of being a major contributor to global warming and other pollution sources. This is mainly because of its heavy use of coal. Of a total of 1.4 billion tons of coal consumed, about 84 per cent is used for direct combustion. In 1999 smoke discharges and dust emissions reached 11.6 Mt, of which 18.5 Mt were sulphur dioxide.[8] Of these amounts, emissions from coal fires represented 70 per cent and 85 per cent respectively. The estimated greenhouse gas carbon dioxide emission caused by energy consumption was 840 Mt of carbon, of which 82.8 per cent was from coal burning (China's Environmental Situation Bulletin quoted in Horii and Gu 2001: 209).

Pressures have increased from international organizations for China to reduce the ratio of coal in her primary energy consumption structure, especially since 1997 when the Kyoto Protocol was agreed. And this is despite the fact that, on a global scale, the degree to which the advanced economies are able collectively to oblige developing countries like the PRC to replace its dependence on coal is very uncertain. The Protocol remains powerless in obliging developed countries to implement it, while developing countries are not subject to the Protocol.

Even without external pressure, internal pressure is growing. Due to the serious pollution in the major cities, the State Environment Protection Bureau has issued very strict rules and regulations[9] on the emission levels of sulphur dioxide and carbon dioxide in 'two control zones' (*liangkongqu*). Large cities such as Beijing have also passed regulations to limit the emission of sulphur dioxide and carbon dioxide These new regulations require not only an improvement in the quality of coal, but also the implementation of clean coal technologies (CCTs) in the whole process from extraction and washing, to burning of the coal and so on. As a further example, coal transport from some provinces, where the coal has high sulphur and ash content, to other provinces has been prohibited.

There is a wide understanding in China that in the pursuit of sustainable development, 'not only must the economy be developed, but also the environment must be protected, so as to have a double victory in both the economy and the environment' (CEDREC 2001: 97). However, as has been shown above, the large-scale use of coal in the country is almost unavoidable due to the lack of any alternative energy source. It is therefore widely believed that it is the direct combustion of coal which should be reduced, rather than the amount of coal consumed in primary energy consumption. This implies increasing the use of coal which has been processed through clean coal technology or converted to a clean energy source. The CCT programmes in China therefore endeavour first to create clean coal while at the same time converting from coal to electrical, oil or gas-powered energy. In 1995, in response to this demand, China set up the Leading National Group for CCT Planning and Implementation, and in 1997 issued *The Development Strategy for CCTs during the Ninth Five-year Plan and by 2010*. However, huge amounts of capital are required to improve and implement domestic CCTs and to import advanced CCTs from outside China.

Washing coal is the preliminary step to reducing the emission of sulphur dixoide and carbon dioxide. However, Table 3.5 shows that the national average rate of coal washing was only 26 per cent in 1997.[10] Facing intense pressure because of environment concerns, the washed-coal ratio was reported to reach 35 per cent in 2000. However, this ratio has to reach 50 per cent by 2005 under regulations from the State Environment Protection Bureau. To reach this goal, 71.5 per cent of China's mines need to build coal-washing facilities; 40 new coal-washing plants with an extra capacity of 100 million tons need to be built during the Tenth Five-year Plan 2001–5.

China's rapid industrialization since the late 1970s has required an equally rapid growth in electricity supply – on an average annual growth rate of 8 per cent. To match this, coal for thermal power plants increased from 164 Mt in 1985 to 643 Mt in 2001. The ratio of coal for electricity generation to total coal output increased from 17.98 per cent in 1980 to 58.20 per cent in 2001 (CEPY 2002: 642). Significant increases in percentage of coal used for electricity generation are planned in the future.

While such a huge amount of coal is consumed in power plants, 70 per cent of 500,000 sets of coal-fired boilers were mainly medium- and small-sized layer-combustion boilers. In the 1980s 100–300 MW units were the mainstay, and 300–600 MW power units more commonly used in the 1990s. Typically these small-scale power plants were much less efficient in converting coal to electricity and had higher levels of emission of pollutants per unit of electricity (World Bank 1997b: 55), and this is indeed a major origin of pollution.

China deployed three strategies to combat this problem. A closure policy for most small coal-fired power plants has been implemented since the late 1990s.[11] Meanwhile, outdated facilities have been replaced. In 1997 $1 billion was spent importing electricity generating equipment, of which $522 million was to import boilers used for generating electricity, up 40 per cent on 1996 (CEY 1998: 250). In addition, the government intends to increase the number of large pithead power plants alongside large coalmines. With the transmission of electricity from the pithead to the place of consumption ('coal by wire'), much of the pollution otherwise caused by the transportation of coal by road or rail can be eliminated.[12] Although all of these three routes will be costly and some may incur resistance, there is little choice for the government when facing higher expectations from both international and domestic environmental groups.

Direct combustion of coal by industry (with exception of power plants) and households also accounts for about half the total coal used. To reduce pollution from this channel, thousands of old-fashioned boilers, heating methods, fans and pumps used in heating and so on, also need to be replaced, as they too contribute to sulphur dioxide and carbon dioxide emission.

Several coal gasification programmes, including both underground gasification and surface gasification, have been implemented in China, but the cost of producing gas through the current domestic technology was over 1 yuan per cubic metre, almost double the current price of gas in the major cities, which is 50–70 cents per cubic metre. Comparatively, coal liquefaction programmes are preferred by the government. However, because of the huge cost as well as the potential risk in the process from applying laboratory technology to industrialization, the central government is still waiting for the outcome of the pioneer coal liquefaction industrialization project in the Shenhua Group before making a decision on whether or not to expand these projects to a large scale.

All these efforts and possible improvements to deal with the challenge from sustainable development will, to a large extent, depend on available funding, and this unfortunately brings us back to the original challenge from development.

Transition

The coal industry has been challenged by the problems created by the system transition vital to the survival and development of the coal industry. Since the command economy system penetrated everywhere, deeply affecting the coal industry in particular, the transformation of this industry became especially problematic. Since the late 1990s the challenges posed by globalization, together with the revolution taking place in global business practice, have aggravated the problems caused by transition.

As demonstrated by China's history of decades of coal shortages, the command economy system has failed to make maximum use of internal funding and production capacity to meet the high demand for coal, and was unable either to deploy a rational price structure to protect the interests of coal producers or to depress unreasonable demand, or to build and organize sufficient transportation capacity to ensure that coal reached the customers on time and in sufficient quantity. All these factors self-evidently demonstrate the need for transition.

Until now the pace of the transition in the industry has been far slower than in other industries. Government guaranteed loans did not end until the late 1980s. Prices were not liberalized until 1993, while those for most other commodities were liberalized much earlier. Subsidies to key SOE coal companies are still allocated, although the amount and its channel have changed. National freight capacity is still allocated at the Annual Coal Convention, where major customers are still required to sign contracts to procure coal from certain suppliers. More seriously, over a period of more than ten years of reform and transition, the coal industry is not better but worse off in terms of its extremely difficult financial situation and huge losses incurred during the transitional period. This implies that transition in this industry is far from complete and huge difficulties still lie ahead.

The need for transition

The strategic importance of coal to China, the history of shortages, and the long distances between coal producers and customers, all determined that the coal industry had to be tightly controlled by central government and under the terms of the command economy system. The question as to why this system needed to be transformed was discussed at great length immediately after the reform. At a coal reform seminar in 1983 attended by all the top officials from the provincial and autonomous region coal administrations and with participation by key coal companies, eight major weaknesses of the coal administration system at that time were highlighted.

These weaknesses were:

1   The planning administration was over-regulated *(guo si)*.
2   The production distribution was over-regulated.
3   The price system was irrational.
4   All of a firm's profits were paid to the central government, while all expenditures and subsidies were paid by the central government *(tongshou tongzhi)*.
5   The labour force was recruited by government and not by the companies themselves.
6   Welfare distribution was equal regardless of contribution.
7   Coal distribution and transportation were blocked by regions, sectors and departments, restricting easy marketing and export of the coal.
8   Management was divided among many people, with each area of responsibility being shared by more than one 'leader' resulting in a lack of a sense of responsibility on the part of any one of them. Constant bickering and arguing *(che pi)* often resulted (CCCIEC 1989:96).

These weaknesses were abundant proof of the need for transition. The following paragraphs highlight the problems caused by the planning, finance and price administrations in order to show the need for transition in more detail.

Planning administration was the central feature of the command economic system. As one of the most important commodities 'under unitary distribution' by the state, coal was controlled tightly by the central authority, mainly the State Planning Commission. Long-term and short-term plans covered all major targets of the coal industry ranging from mine construction and procurement, to output, distribution, transportation, investment and pricing. Above all was a comprehensive system planning the supply of, and demand for, raw materials whereby intermediate goods and finished goods all had to balance exactly. It was planners rather than enterprises themselves who 'calculated how for every state enterprise these were derived, and how and when they were to be distributed' (Thomson 2003: 27).

This planning system was positive in that it could enable the new government after 1949 to meet the demand from the most important industries such as steel and defence and thereby fully control the national economy.

However, the crucial problem of such a system was that '[I]f either the plan or its execution were faulty, cumulative dislocation occurred' (Thomson 2003: 27). Unfortunately, during the entire command economy period, state planning was no more than a political slogan driving the industry to pursue higher production. It failed to provide guidance for industry as it changed frequently in order to adapt to political need rather than the practicable capacity of the industry. Serious consequences of this included the consistent imbalance between production and preparation, and huge losses in funding, coal resources and human resources that were common among 'soon to start, soon to close' construction projects.

This planning system also led to the separation of production from sale and transportation. During most of the period under the command economy, coal from key SOEs was mainly distributed unitarily by the state.[13] At the same time coal under unitary distribution was also allocated transportation quotas to guarantee its transportation. During the years from 1949 to 1951 and from 1978 to 1981 sale departments, independent of coal companies, were responsible for distributing coal 'under unitary plan' to planned customers or selling the surplus to market.[14] While coal companies had little knowledge of the requirements from sellers and customers, these sales departments did not provide advice to producers. Disagreement between producers and sellers therefore often occurred in the process of examining coal for quality, handing over and taking over of the coal, and storage and inventory checks (CCCIEC 1989: 561). Although, with the exception of these years, responsibility for sales was returned to coal companies and consequently coal producers were better able to fulfil distribution and transportation reguirements, SOE coal companies still did not need to have a clear understanding of the market or their customers. Because coal was in severe short supply, SOEs were not concerned as to whether or not their coal would be sold or distributed. This situation, among others, enables one to understand why SOE coal companies were so uncompetitive *vis-à-vis* TVE mines. The latter were much more familiar with their markets than SOEs, as they had a much longer history of selling their own coal.

Key SOEs were allocated quotas for coal to be transported under the command economy, but this was not without problems. Due to the huge amount of coal to be transported, the long distances which separated coal customers from producers, and especially the severe shortage of rail capacity, accessing this was always a difficult task for coal producers. It required that the three separate plans for production, distribution and rail transportation were well designed and fully executed in a cooperative way. However, under the command economy these three plans were never properly integrated. At each Annual Coal Convention, producers, transporters and customers were in continual dispute. The normal situation was that the distribution plan exceeded the production plan, and the production plan exceeded the transportation plan. The consequence was that the transportation plan had to be cut month by month, leading to a situation whereby every month plans had to be modified, and every month these plans failed to balance one another. This greatly affected the balance between supply and demand leading to a chronic and chaotic mismatch among the needs of the three components (CCCIEC 1989: 567).

Although key SOEs historically enjoyed the transportation quotas at Annual Coal Conventions, and this continues to the present day, they face difficulties from several sides. One difficulty is that the guaranteed quotas are decreasing and SOE management has to compete for them with other coal producers and rail users. Another difficulty stems from the fact that transportation costs are rising more rapidly than are the prices paid for coal.

Finance administration was another problem. Under the system of *tongshou tongzhi* (whereby all profits were remitted to central government and all expenditure/investment was allocated by central government), almost all investment in SOEs came from government. One of the serious consequences of such system in the coal industry was that of insufficient investment. Coal shortages were, of course, related to the rapid increase in demand, but were also caused by insufficient supply which in turn resulted from dependence on a single source of investment. By 1980, over 82 per cent of coal output was still funded by central and local governments and only 18 per cent was funded by communes and brigades (Ye and Zhang 1998: 29–30). Although, since 1949, coal was the major focus of government finance, and received larger investment than other industries, the total was never sufficient to meet demand. After the Cultural Revolution especially, when 'everything needs to be done' (*baifei daixin*), the government seemed less and less able to provide sufficient priority funding for the industry. Figure 2.3 shows that the ratios of government capital construction investment in coal to energy industry, to industry as a whole, and to the nation as a whole, all sharply declined since the Sixth Five-year Plan (1981–5). It also should be noted that because the government was investing heavily in the energy industry, people in general did not understand the disadvantage for the coal industry. In fact the majority of investment in the energy sector went to the power generation and oil industries rather than to coal.

These features in particular necessitated the transition of this investment system. During the Sixth Five-year Plan (1981–5), embarrassed that the 'economy wants coal, but government lacks money' (*ye yao mei, er mei qian*), the government desperately encouraged the development of TVE and private coalmines by using social and individual funds. This was known as 'walking on two legs', and proved extremely successful. (Ye and Zhang 1998; CCEC, 1999).[15]

Prices being set unreasonably low was another typical feature of the command economy system. For decades the government had implemented a low price system for coal because of the extreme importance of coal for the economy and for the

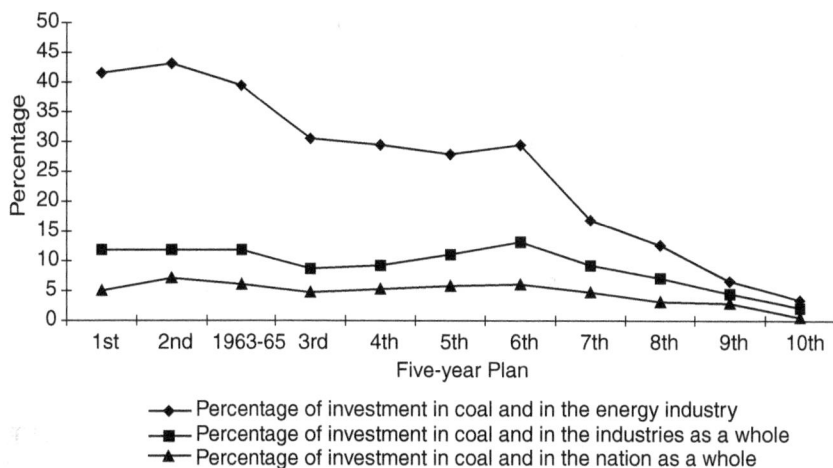

Figure 2.3 Changes in the share of capital construction investment between the coal industry, the energy industry, industry as a whole, and the nation as a whole 1953–2002

Sources: SSB 1985, 1995; CCIY 1999; Thomson 2003: 79–80, 228; CSA 2003: 57.

daily life of the people. For almost all the period between 1949 and 1993 the price of production was in fact much higher than the actual cost of production. This inhibited producers, preventing them from making a profit and in turn reinvesting funds to improve their productivity. Moreover, due to this low price, coal was under-represented in the cost structures of dependent commodities and this in turn contributed to an over-demand for coal and to its inefficient use. Coal was therefore in short supply. This was also the reason that forced the government to increase its subsidies to the coal industry. From 1949 to 1985 the government's total investment in this sector was 75.89 billion yuan. The ratio of profits and tax to investment was a mere 29.6 per cent – much lower than the average ratio of 208.15 per cent for the whole of the industrial sector, and also lower than the average ratio of 92.13 per cent for the mining sector (CCEC 1999).

Well before 1993 when coal prices were finally freed up, proposals to increase the price were common (Ministry of Energy 1991, 1992). However, because of the special position enjoyed by coal in the economy, it was important to minimize the effect of price changes on other (coal-consuming) industries and society as a whole. To encourage competition resulting in improved coal productivity and to minimize the negative impact of dramatic coal price liberalization, a dual-track price system was adopted in the coal sector from 1985.

Under the constraints of the planning administration, by the segregation of production, transportation and sale, and the system of low prices, the SOE mining companies were limited to production. They, therefore, had no need to concern themselves with markets, prices, customers or transportation. In addition to these constraints, wage and welfare distribution was also problematic as it was equal for everyone regardless of their differing abilities and contributions. For these reasons SOE companies fundamentally lacked positive incentives to increase their productivity. Almost the only investment channel, stemming from the central government, was also inadequate to ensure all the necessary funding for this beleaguered industry. Consequently out-of-date technology and equipment was never tackled in SOE mines. Shortages became inevitable. Losses were overwhelming. From 1980 onwards change became unavoidable.

The difficulties of transition

As local SOEs and TVEs were constrained to a lesser extent, or not at all, by the limitations of the command economy system, the need for transition and the challenges from transition applied mainly to key SOE coal companies. Because of their complicated structure and limited by irrational requirements, the transformation of SOEs was both urgent and inevitably problematic.

The transition was actually never more urgent than before the boom of TVE coalmines which eventually led to surplus in the coal market. Under the command economy system, and even during the long period after the reform from 1978 to the mid-1990s, SOE mining companies were not considered to be problematic since they did not need to be concerned in making profits and their losses were balanced by subsidies. However, in competition with thousands of TVE mines and a surplus market, the losses incurred by SOE companies began to accelerate alarmingly as the result of hugely increased costs and rising welfare expenditure, vitiated by their inherent lack of incentive in the search for markets. At the same time, moreover, government subsidies were greatly reduced as a more stringent budgetary policy became effective.

The gradualist transition that has been deployed since the 1980s will be examined in Chapter 4. First, SOEs were forced to become market orientated because TVE coalmines had been introduced into the market. Second, formerly whenever funding was insufficient, SOEs would request more funding from central government, but now these soft budgets were tightened. Third, the 'contract responsibility system' (CRS) was implemented so as to give more management autonomy. Fourth, a dual-track coal price system was applied so as to stimulate the output from SOEs and increase their profits. Finally, when coal was in sufficient supply by the late 1990s, the government started to accelerate the transition by first abolishing the Ministry of Coal Industry in 1998; by decentralizing all key SOE coalmines to local provincial governments; by selling small SOE coalmines; and by bankrupting hopelessly uneconomic key SOE bureaux.

Huge difficulties are emerging as the result of this transition. These changes have radically challenged the privileges and legitimacy traditionally enjoyed by SOE coal workers from the 1950s to the 1990s. Under Mao's regime, SOE workers were known as 'the most glorious class' and granted privileges. Since 1954, and in compliance with this policy, the wage level of the workers in the coal industry, along with those for the oil and ferrous metal industries, was the highest category in all the industrial sectors (CCCIEC 1989: 607).

With the progress of the general economic reform, however, wage levels for workers in the coal industry have been downgraded by comparison with those of workers in other sectors, and are considered among the lowest at the beginning of the twenty-first century. The coal industry has also become more and more disadvantaged since the reform as coal prices were liberalized much later than those of other commodities and they also increased much more slowly than the price of other commodities. Most SOE coalmines were now loss makers, and since 1994 over one million of their workers were dismissed. It will certainly take a considerable time for the labour force of SOE coalmines to come to terms with such dramatic changes.

Transition also challenges the government's financial capability and the former socialist welfare system. China's coal industry was labour intensive with over four million SOE employees by the mid-1990s. It was reported that about one million employees were to be dismissed from the SOE coal sector during the Ninth Five-year Plan (1996–2000), to reach the target of reducing the total employees of SOE coal companies to about three million (SCA 1999a). However, this was only partially completed. This was basically because the costs of dismissing workers or bankruptcy were too high (CCC 2000).

These challenges indicated that the dramatic transformation of SOE coal companies at the beginning of the reform would probably lead to a drop in coal output, causing a severe energy shortage and in turn seriously damaging China's burgeoning industrial sector. It might also have an overwhelmingly negative impact on SOE workers. Because the objective for transition was to serve the ultimate goal of development, a more gradual approach to transition was therefore deemed more suitable for China's coal industry.

Globalization

Despite large oil discoveries in the Middle East in the twentieth century, coal remains one of the most important sources of primary energy in the world. In 2002 coal still accounted for 23 per cent of the world's primary energy consumption, and 39 per cent of electricity fuel, and 70 per cent of global steel production depends on coal feedstock (WCI 2003). Known world coal reserves are thought likely to last for 250 years at current levels of consumption, much longer than the life expectancy of oil reserves. Because of the lack of alternative energy sources, coal is more important for China than for almost any other country.

Coal is also the major fuel used in generating electricity worldwide. Many advanced countries such as Australia and the USA are heavily dependent on coal for electricity, with 84 per cent and 56 per cent of coal-fired electricity respectively. The International Energy Agency (IEA) forecasts suggest that the worldwide use of coal for electricity generation will grow significantly over the next 20 years, particularly in China and many of the developing countries. Coal will remain the most favoured fuel where gas is unavailable or expensive (e.g. China and India). The US Department of Energy and the UK Department of Trade and Industry have made similar forecasts.

During the big business revolution truly global coal giants also emerged, as shown in Chapter 5. Despite the hesitant attitude of most Chinese coal companies, which could be described as 'half welcoming and half resistant', all of the global mining giants have entered China and have started to challenge this most traditional industry, which seems apparently unprepared to meet their challenge.

China was the first country in the world to discover, explore, and use coal (CCCIEC 1989: 1). Because of its long history and the relatively low technology involved, China has always appeared to have no urgent need to cooperate with the outside world. After 1972 the closed-door policy in the coal industry was changed and China began importing advanced equipment, and this movement has expanded since 1978 (CCCIEC 1989: 109).

It is interesting to note that while both the coal and oil industries are energy industries and both are considered strategic, the government's attitude to foreign investment in them is totally different. From 1982 China's offshore oil corporation started to cooperate with foreign companies. By 1999 it had signed 142 contracts with 69 foreign companies in 18 countries, with absorbed foreign direct investment (FDI) of over $6 billion. Meanwhile, China's onshore oil companies had signed 56 contracts and absorbed FDI of over $1.1 billion. Thus, the FDI in the oil industry by 1999 was over $7.1 billion, which excluded foreign loans believed to be at least as much as the FDI. Joint ventures in the oil industry are actively operating their business in all of China (CEYEC 1998: 157). In contrast, during the two decades of the reform period from 1978 to 2000, the total foreign investment in China's coal industry, including foreign government loans, world financial organization loans, commercial loans and FDI, was only $4.17 billion (CCC 2000: 57).[16]

There are only two major joint ventures in coal production: the former Antaibao mine[17] and the current Asian American Coal Ltd. While the former eventually pulled out of the project in 1991, the latter is small scale and still not fully operational. Asian American's interests are not only in coal, but in coke and coalbed methane (CBM) in China. The world's largest three, BHP-Billiton, Rio Tinto and Anglo American have at least carried out negotiations and detailed studies over a period of three to five years, but all admit to not having found suitable partners in China.

The reasons for the limited foreign direct investment in the Chinese coal industry include a general reason due to the nature of the mining business, and a specific reason due to China's investment environment. With regard to the nature of the mining industry, it is a highly capital-intensive industry, requiring long-term investments and a huge infrastructure. This makes the provision of a stable investment environment and secure profit margins in the mining industry much more important than in other industries. It is, however, believed that the conditions for investment in China's mining industry are not ideal (Rio Tinto China 2001, 2003; Asian American Coal 2001; Coaltrans China 2004).[18]

The following six factors are fundamental to understanding the problems of the investment environment for mine exploration in China:

1   The central government's ambiguous policy and the absence of any clear procedures to attract foreign investment in the coal industry.
2   The huge difference of policy among local governments on one hand and the differences in practice between local and central governments on the other,

resulting in obliging potential foreign investors to study not only China's coal market in general but also local market policy and regulations – including an understanding of underhand and corrupt practice.

3   A proportion of the laws and regulations governing foreign investment on coal are still not in place.

4   The peculiar characteristic of the Chinese coal industry in which producers are separated from their markets on average by distances of over 550 km resulting in considerable transportation problems. This not only increases costs, but also concerns for easy access to the relatively limited number of rail networks which give preference to SOE coalmines over other producers.

5   The extremely poor safety record, raising fears that an accident in a mine in which the foreign company was involved would damage that company's international reputation, which may have been built over decades.

6   The fact of the existence of large-scale povery in China determined that TVE coalmines still continue to operate and enjoy a not inconsiderable share of the market. International mining giants are unlikely to be able to share that low-profit market, but will certainly be affected by it as it will easily take the market into surplus.

But for all that, and in spite of these disadvantages, the general opinion of all these global giants is that the coal market in the Peoples' Republic of China is simply too big to be ignored. Besides, it must also be noted that to open its door to face the challenge from globalization is also the positive action of the Chinese coal industry.

Opening doors to meet the shortage

China's coal industry opened its doors to the world in 1972, much earlier than the nation's formal open-door policy launched in 1979. Constrained by low productivity and severe shortages, China had to make every effort to improve coal production and to do so required advanced equipment and technology.

While the entire machine industry lags far behind world leaders, China's coal mining equipment and management was said to lag at least 20 years behind. While advanced countries now aim to use automated production and management systems, China's non-mechanized coal production accounts for over 60 per cent of the total. While domestically produced coal extraction equipment must be serviced after 1 Mt of produduction, imported equipment is inspected after more than 10 Mt. As a result, almost all of the advanced mining equipment in China's coalmines is imported.

From the 1950s to the 1980s about four large equipment import records were set: (1) in 1973, under premier Zhou Enlai's supervision, China imported 43 sets of comprehensive cutting equipment from Germany, Britain and Poland, valued at $88 million; (2) in 1977, 12 sets of comprehensive cutting equipment were imported from the USSR and 13 sets from Poland, and a further 13 sets from Poland in 1980; (3) in 1978, under Deng Xiaoping's supervision the largest import of 100 sets of comprehensive cutting equipment from Germany, Britain, France and Japan was completed, valued at $730 million – at the same time 100 sets of shaft preparation equipment were also imported; and (4) in 1984 45 sets of comprehensive cutting equipment were imported from Poland, valued at 300 million Swiss francs (CCEC 1999: 265–80).

During the Ninth Five-year Plan (1996–2000), total FDI in the coal industry was $1.14 billion, of which $480 million was used to import coal-extraction and transport equipment. Taking China's most advanced and competitive coal company Shenhua as an example, in the 1990s a single Shenhua project had imported 1.2 billion yuan of foreign equipment. Shenhua Group's best coalmine – Daliuta mine – had 98 per cent of its equipment imported from the USA, UK, Germany, South Africa and other countries.

In the face of closer competition from global mining giants after accession to the WTO, China has to import more to enable its best coalmines to reach their maximum capacity. As a result, numerous mine equipment manufacturing or component companies, such as Joy Mining from the USA and Debert from Germany, have set up business centres in major cities to provide pre- and after-sales services. Their business is much more active than companies purely producing coal.

Huge potential attracts the entry of multinationals

Besides China's vast coal business market, her coal-related business is equally large. This at least includes the market connected to the implementation of CCT, coal liquefaction and gasification, mining equipment, mining and shaft design co-operation, coal washing equipment, coalbed methane exploration and development, and other mining product businesses such as the aluminium and the coke businesses. It is indeed a market which is too large to be ignored.

Besides the huge import market for mining equipment, implementing CCTs is another vast market. The worldwide controls on the emission of greenhouse gases, sulphur dioxide and nitrogen oxides provide a huge market for developed countries to export both their technology and relevant CCT equipment to countries in need of it. It would be 'a bonanza for companies in the energy efficiency business' (*FT* 28 November 1998). The UK Department of Trade and Industry forecasts the value of the global CCT market could be up to £300 billion for the decade 2001–10, and much more beyond 2010. Of which UK companies could benefit to the tune of some £30 billion between now and 2010 and hope to establish a significant foot-hold for the longer term. Because of the degree of pollution resulting from coal use, as the analysis above shows, China is obviously one of the biggest CCT markets.

Moreover, some large coal companies, such as Shenhua, have already grown to be able to compete with foreign coal companies. Entry to China could place foreign companies in a better position to deal with China's coal producers, such as Shenhua, and undermine them as competitors in the world coal market.

China's coal industry is not ready for the challenge

The challenge from the trend to globalization is reflected in the widening gap between China's coal companies and their counterparts elsewhere in the world. The competitive advantage in the global coal industry lies primarily in size, experience and management skills, better equipment, technology, productivity and brand name. Their Chinese counterparts are uncompetitive in all of these areas.

As shown before, there are no economies of scope and scale at all. The largest coal producer, the Shenhua Group, accounted for only 5 per cent of the national total, so the scale is small. In terms of scope, the coal industry has for a long time been focusing on coal business only, while related profitable or high-value-added products such as coke, chemical products and fertilizers are all allocated to other ministries or departments. Even the coal companies' very limited electricity generation from pithead power plants was squeezed out of the electricity market by not being allowed to access the major grids. SOEs are suffering financial difficulties and TVEs are being closed. The productivity for China's SOE coalmines was 1.9 tons per employee per day, equivalent to two per cent of that in the USA. Even for modern coal corporations, such as Shenhua, there are still many constraints on their capacity to compete with the international mining giants. Among these constraints are the obligation to look after loss-making SOE coalmines, insufficient capital funds, and the heavy social welfare burden.

Joining the WTO has simply turned these previously remote threats from the global giants into direct challenges. For example, foreign coal companies will find it easier to establish their own or joint-venture companies to compete directly for China's huge domestic market. Domestic coal customers – especially in the south China provinces – may import more coal and thereby reduce domestic coal demand. China may import more products which use coal, such as steel and construction materials, and so reduce domestic coal demand. More environmental pressure will be imposed on China's coal production and consumption.[19]

The challenge from globalization in the coal industry is centred on the difficulties in building modern firms. As will be shown in Chapter 5, the difficulties in building modern coal corporations on a large scale in China exist in every aspect, from continued public ownership, insufficient finance, insufficient expertise in combining both the culture of Chinese firms with modern coal corporation governance to the constrain of diseconomies of scale, and the difficulty of organizing TVE coalmines and loss-making SOE coalmines. These challenges determine that China must use its industrial policy to strengthen indigenous firms to meet the impending challenges. Without government support and coordination, it may be too difficult for her small and weak coal companies to be competitive in the global market.

Inter-relationship of the three challenges and the role of the state

China's coal industry has been facing the challenge from development, which is reflected in the fact that coal production was exhaustively pursued to fuel the economy, to overcome the shortage of supply, to eradicate poverty, and to ensure sustainable development.

It was this challenge, posed by the needs of development, that called for a system change which enabled the coal industry to benefit from the development of TVE mines, and the introduction of the open-door policy which attracted assistance from countries with more technologically advanced mining experience. Such a system transition from one where the state dominated to one involving the combination of both planning, market supply and the import of advanced equipment and technology, has dramatically improved coal output, ended the long history of coal shortages, ensured the boom in the economy, and improved living standards among the rural population. In addition, both the pressures exerted by international influence and the technological assistance that it brought with it may also contribute to the sustainable development of the Peoples' Republic.

While the coal industry in China benefited from these positive and mutual effects among the three challenges, it also had to deal with their possible negative relationships: the development of TVE coalmines and the resulting surplus coal market led to the worsened financial situation of the SOE coalmines and forced them to restructure or go bankrupt, and constrained the integration of the coal market and the building of modern coal corporations such as Shenhua.

Meanwhile, because of the difficulties in transforming the SOE mines or make them bankrupt, the government had to close more TVE coalmines to reduce surpluses and allow SOE coalmines to continue functioning for longer, and promote the establishment of modern coal corporations.

Central government strove to deal with each of the three challenges, but was extremely embarrassed in dealing with the complicated relationship between the three challenges. In the long term, Shenhua is the hope for China's coal industry. However, the extent to which more Shenhua-style coal corporations can be built will largely depend on China's development and its transitional progress: the government's financial capacity to build more 'Shenhuas'; the speed at which a mature social security system can be established to resettle redundant workers so that more labour can be replaced in SOE mines; and the real number of 'closed' TVE coalmines. Obviously, all these uncertainties will require profound government involvement.

Conclusion

China's coal industry has to meet the three challenges. It has to ensure sufficient supply to fuel the nation's economic growth, to promote rural development, and to pursue sustainable development by sharing obligations to reduce pollution which results from the use of coal. However, the coal industry faced extreme difficulties in carrying out these tasks under the command economy system. Transition and an open door policy were called for. The system transition, including allowing the development of TVE coalmines and the reformation of SOE coalmines, successfully ended the history of coal shortage and promoted rural development. It has, though, proved very difficult. Meanwhile, the coal industry also opened its doors much earlier than other industries, to import advanced equipment and technology, and to improve productivity. However, against a background of globalization, and attracted by China's huge coal market, all global mining giants have entered China. The fragmented coal industry is therefore facing more intense threats from them.

As a result, the Chinese government has been facing a dilemma in dealing with the interaction between these three challenges. To pursue development, permission was given for the TVE mines to operate longer, which resulted in the SOE mines becoming uncompetitive and being forced to transform. Because a change of this sort entailed so many problems, the government had to resort to the closure of the TVE mines so as to keep the SOE mines functioning. By pursuing development, an open-door policy became necessary. However, as globalization now challenges tens of thousands of Chinese coal companies, the government resorts to building large coal-producing corporations such as the Shenhua Group, and in doing so will undoubtedly bring about the integration of the coal industry, but at the cost of making many more workers redundant, closing more TVE mines and bankrupting more SOEs. The struggle between the three challenges is inevitable and the state has to deal with these conflicting issues in a well-balanced and cautious manner.

3 Development

The rise and fall of TVE coalmines

Poverty in the LDCs was and still is mainly rural. ... For many years, until about the 1970s, rural development was widely neglected. ... Rather than being permanently immersed in a subsistence economy, the rural poor are increasingly dependent on incomes derived from marketing their labour, often in nonagriculture activities. These developments, which are especially powerful in China, stress the need for a more diversified approach to rural development.

(Hogendorn 1996: 348)

China has 80 per cent of its population living in the rural areas. Whether China could maintain stability and whether China's economy could be developed depend on whether the rural areas could be developed and whether peasants' life could be better.

(Deng Xiaoping 1984: 77–8)

In the primary period of socialism, to develop TVE coalmines is an unavoidable choice [for China]. Our country has large reserves and location varied coal resources, and also has large amounts of surplus rural labour force. These are two major factors of the development of TVE coalmines. Meanwhile, the high demand for energy provides the original push for TVE coalmines' emerging and development.

(Ye and Zhang 1998: 16)

Introduction

In China, township- and village-owned enterprises (TVEs) (*xiangzheng qiye*) include not only collective, but also private and individually owned enterprises (*jiti qiye, siyou qiye he geti qiye*).[1] The rapid expansion and great commercial progress of township- and village-owned enterprises in the Chinese economy have been praised as one of the most remarkable features of her economic development in the last twenty years (Nolan 1988; Nolan and Dong 1989; Byrd and Lin 1990; Ho 1994, 1995; Weizman and Xu 1994; Luo *et al.* 1998, 1999; Jin and Qian 1998; Lin and Yao 1999). Table 3.1 shows that TVEs, in general, do merit such praise.

The TVE 'economic miracle' led to many claims that China's success depended on the flourishing of TVEs. While large state-owned enterprises (SOEs) were assumed to be declining or dying, TVEs were therefore expected to be the major source of China's future development (World Bank 1996; Sachs and Woo 1997; Pack 2000).

Since the late 1990s, however, the Chinese government has implemented a large-scale closure policy for TVEs, from coalmines to power plants, from cement factories to fertilizer factories. The basic reason for this has been a number of significant negative impacts of TVE development. By presenting a picture of stagnating TVEs which, since 1996, were absorbing fewer unemployed workers, the Department of TVEs in the Ministry of Agriculture expressed considerable concern about the fact that the slowdown of TVEs might not enable the government to achieve its goal of TVEs absorbing an additional annual average 2.5 million rural labour force during the Tenth Five-year Plan (CEY 2002:708).

This raises a number of questions. If the negative effects of TVEs have been so great, why were they so strongly encouraged to develop in the first place? And if the TVEs' contribution is really as great as anti-gradualists believe, why has the Chinese government insisted so firmly on closing them down; and why has such a large proportion of them been closed since the late 1990s?

This chapter, based on detailed case studies of several TVE coalmines in northeast, north-west and central China in particular, and the entire TVE coalmines and the coal industry in general, aims to present an unbiased and comprehensive picture of China's TVEs, to establish the causes of their rise and fall, and to assess their proper position and treatment in China's economic development. The second section provides an overview of the rise and fall of TVE coalmines; the third section explores the major forces behind the dramatic development of TVE coalmines; the fourth and fifth sections highlight TVE coalmines' great contributions and equally great negative effects respectively on China's coal industry, economy and environment; the sixth section examines the causes of closure of TVE coalmines and the reasons for resistance to this closure. Finally, a

Table 3 1 The great achievement of TVEs 1978–96

	1978	*1996*
No. of TVEs	1.52 million	23.36 million
No. of employees	28.27 million	135.08 million
Share of rural labour force	9.5 %	29.8 %
Share of rural income	7.5 %	34.2 %
Share of gross rural output	21.2 %	77.2 % (1995 data)
Share of industrial output	9.1 %	57.9 % (1997 data)
Share of exports	9.2 % (1986 data)	45.8 % (1997 data)

Sources: CTVEYEC; SSB, various years; Lin and Yao 1999; CSD 2003: 42.

Note
1996 was selected, rather than 2002, for comparison with 1978, because 1996 might be regarded as the peak for TVE development. For example, by 1999 the number of TVEs had decreased to 20.71 million, and employees to 120 million. By 2002, employees were 132.88 million, which was still fewer than that in 1996.

conclusion is finally reached about TVE coalmines' rise and fall. Challenged by the underdevelopment – typically vast population versus limited natural resources, rural poverty and large underemployment – China must make maximum use of TVEs' contribution. However, challenged by transition and globalization – basically the requirement of industrial restructuring – China must restructure and even 'kill' TVE coalmines.

Overview: the rise and fall of TVE coalmines in China

TVE coalmines are defined as 'those coalmines collectively owned or operated by towns and villages, those privately owned, and those excluded from state-owned coalmines and foreign investment coalmines' (Ye and Zhang 1998: 16). Despite the many categories, the Chinese in fact popularly call all of them TVE coalmines (*xiangzhen meikuang*) or small coalmines (*xiao meikuang*).

Small private coalmines existed well before the Chinese Communist Party (CCP) established the People's Republic of China (PRC) in 1949.[2] Under Mao's regime these previously small private coalmines were restructured partly as community and brigade enterprises (CBEs) and partly as local state-owned coalmines. To be in line with the popular slogans of 'exceeding the UK, catching the USA' (*chaoying ganmei*) and 'doubling steel output' during the Great Leap Forward (1957–9), the Ministry of Coal Industry called for 'all the people to open coalmines'. The number of CBE coalmines was over 100,000 and they employed over 10 million peasants to extract coal (Ye and Zhang 1998: 25).

However, most of these enterprises were closed during the restructuring period from 1958 to 1966. During the politically chaotic period from 1966 to 1976, though in principle 'the countryside is generally not allowed to operate industry' (Ye and Zhang 1998: 17), TVE coalmines were exceptionally encouraged from 1969 to meet the high demand for coal. However, due to the large-scale closure from 1958 to 1966 and the cessation of production due to violence at the beginning of the Cultural Revolution from 1966 to 1968, by 1978 the number of CBE coalmines had decreased to 17,800, the output was only 95 Mt, 15.4 per cent of the national total (Figure 3.1, also see Table 2.2).

From 1978, however, mainly because of a severe coal shortage and Deng's regime having shifted the rural development strategy, TVE coalmines developed dramatically. At the peak year 1996 the number of TVE mines rose as high as 63,975, output as high as 638 Mt (Figure 3.2), 46 per cent of the national total. This not only turned the coal market from severe shortage into surplus to support fast economic growth, but also dramatically changed the coal industry's ownership structure from one in which SOEs accounted for over 80 per cent before 1979 to one in which SOE and TVE coalmines each comprised half of the national output (Figure 3.3).

TVE coalmines development since the economic reform can be divided into five periods: an initial period (1979–82) and an adjustment period (1989–91), two periods of dramatic rise (1983–8 and 1992–8) and a dramatic decline (since 1998).

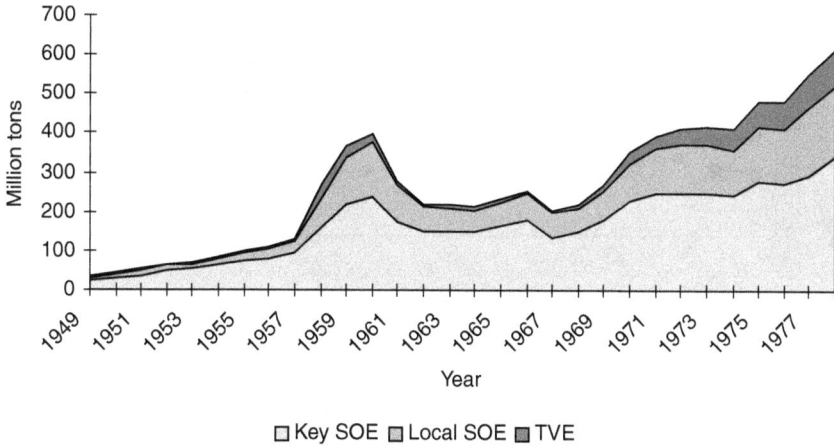

Figure 3.1 Coal output in China, 1949–78 (Mt)

Sources: SPC 1997: 131; CCIY 1999; SCA 2000: 2; 2001: 1.

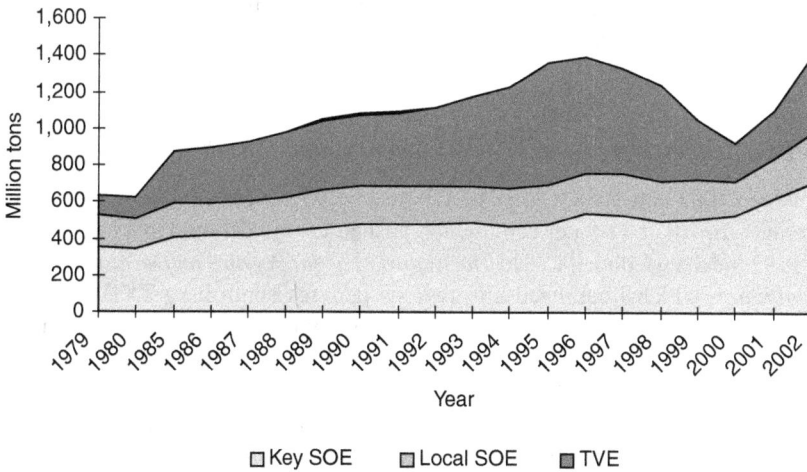

Figure 3.2 Coal output in China, 1979–2002 (Mt)

Sources: SPC 1997: 131; CCIY 1999; SCA 2000: 2, 2001: 1.

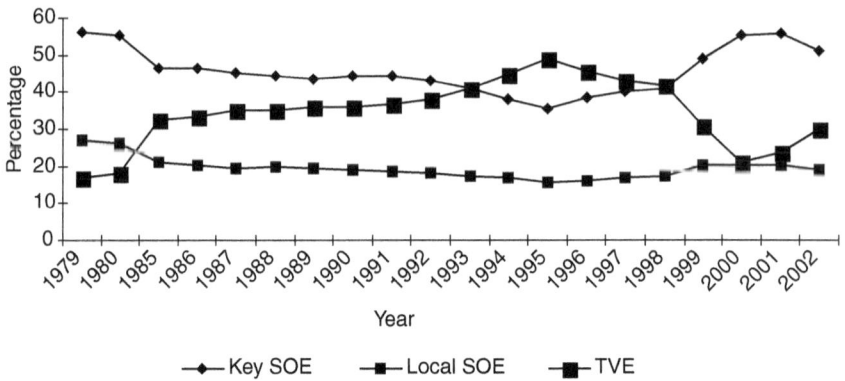

Figure 3.3 Changes in production ratios between producers 1979–2002

Sources: SPC 1997: 131; CCIY 1999; SCA 2000: 2, 2001: 1.

TVE coalmines' initial development period 1979–82

In 1981 the Chinese government issued an eight-word policy to explicitly support TVE coalmines development: *zhi chi, tiao zhen, gai ge, lian he* ('support, restructure, reform, and integrate'). With this policy, output from TVEs increased from 106 Mt in 1979 to 146 Mt in 1982, an annual growth rate of 11.2 per cent. The specific feature of this development period was that these coalmines developed quickly but in a controlled way under, for example, official registration (Ye and Zhang 1998: 30–1).

TVE coalmines' first dramatic development period 1983–8

Between 1982 and 1988, China's GDP rose by 95 per cent, recording an annual average growth of 11.8 per cent, which pushed energy demand to a higher level. In 1983 the State Council issued *The Report of Eight Policies on Developing Small Coalmines*, which liberalized all previous policies controlling TVE coalmines development: encouraging private firms or individual people to raise funds to operate coalmines, awarding autonomy on coal output, sale and transport, acting to resolve the shortage of production materials, offering tax holidays and even some special lower interest rate loans from local finance (*nongchun xinyong hezuoshe*). In 1984 *The Announcement on Further Policy Relaxation and Open Mind on Local Coalmine Development* made TVE policy clearer: 'to mobilize all possible sources, to raise all possible funds, to assemble all large, medium and small coalmines; to pool the efforts of state, collectives and individuals [to meet China's coal shortage]' (Ye and Zhang 1998: 33). Most importantly, in 1984, the then top leaders Zhao Ziyang and Hu Yaobang visited coalfields in western China and exhorted local peasants by using the slogan 'let stagnant water flow quickly' (*youshui kuailiu*), with an explicit meaning of extracting coal in as much quantity and as fast as possible. After that, the slogan was advocated throughout the whole of China, and TVE coalmines developed explosively.

Consequently from 1982 to 1988, TVEs coal output rose 140 per cent from 146 Mt to 351 Mt, at an average annual increase rate of 15.8 per cent, 4.5 per cent higher than that of the previous period from 1979 to 1982. By the end of 1985, the number of small coalmines was up to 63,000, of which 50.9 per cent did not obtain official licences, including the earlier single 'mining extracting licence' (*kuangshan kaicai xuke zhen*) and the later 'coal business operation licence' (*shengchan xuke zhen*).[3] Most of the illegal coalmines started up after 1984. This was the beginning of the disordered period of TVE coalmine development (Ye and Zhang 1998: 35).

TVE coalmines' first restructuring period 1989–91

China's overheated economy was in downturn during this period, and there followed the issue of the policy of 'dealing with economic environment, restructuring economic order, comprehensively deepening reform' in September 1988. China's GDP growth rate slowed from 11.3 per cent in 1988 to 4.2 per cent in 1990, with the result that coal demand was held back. This enabled the government to have a breathing space to deal with the problems resulting from the dramatic development of TVE coalmines during the previous period. After the restructuring, the number of TVE coalmines was reduced from 91,000 in 1989 to 68,000 in 1990. Meanwhile, the safety record improved and the average output of mines increased.

TVE coalmines' second dramatic development period 1992–8

Deng Xiaoping's southern tour in 1992 initiated another strong upturn in China's economy. GDP in 1995 was 60 per cent higher than that in 1991, an annual growth rate of 12.5 per cent. While coal was still in high demand, the government emphasized that the formal policy 'to assemble all large, medium and small coalmines to meet China's coal shortage' would not be changed. However, being aware of more problems in the TVE coalmines, the specific policy on these mines was changed to the Chinese phrase *fu chi, gai zhao, zheng dun, lian he, ti gao* ('assist, restructure, adjust, combine and improve'). This indicates that, despite the problems of the mines, the policy was still to improve them. Reform and adjustment of TVE coalmines was seen as the way forward. To develop the coalmines, the government even issued a 500 million yuan special loan to be spent on reforming them.

Consequently, from 1992 to 1995, the output of TVE coalmines increased at a very high rate: from 400 Mt in 1991 to 659 Mt in 1995. The increased output of 259 Mt accounted for 93.5 per cent of the total national increase in output from 1991 to 1995. This was the peak period of output growth for these mines. From 1995 to 1998, although the output still rose, the rate of growth declined, basically because the coal market had turned to surplus (Ye and Zhang 1998: 41).

TVE coalmines' dramatic period of decline from 1998 to 2001 and the setback afterwards

Since 1998 the Chinese government has been accelerating the implementation of policy called 'closing mines, depressing output' (*guanjin yachan*). Although SOE coalmines are also required to reduce output and dozens of SOE coalmines have been bankrupted, this policy focuses mainly on closing TVE mines and limiting their production.

Although the principal policy for TVE mines is still to suppress their wild development, the policy has been more flexible since 2002 when coal shortage was consequently becoming a problem again. Currently (2004) the government emphasizes quality rather than quantity of TVE mines. In other words, TVE mines which can meet the safety requirements and have all necessary licences are encouraged to develop.

The major forces for the development of TVE coalmines

While free-market 'fundamentalists' recommended privatizing large and medium SOEs to pursue the transition, China embarked on a typically gradualist reform which involved the fast development of non-state-owned producers – mostly TVEs – to compete with non-privatized SOEs.

Four factors to determine the rise of TVE coalmines

Four fundamentally inter-related factors determined the mushrooming of TVE coalmines in the early 1980s. First, China was constrained by severe coal shortage as a result of both prolonged accumulation and the rapid economic growth after the reform. The state was starved of investment funds and could not provide the necessary investment in SOE coalmines. Second, the rural population had a strong desire to escape poverty and improve their living standards by whatever means they could. Third, coal-rich rural areas also had a large surplus labour force. Finally, a series of government policies, recognizing the favourable reasons for TVE development, promoted TVE coalmines. One of the most important policy changes was liberalizing the former restrictions on mobility of the population and labour force in general, and issuing a new policy of 'leaving the land but not the countryside, entering the factory but not the city (*li tu bu li xiang, jin chang bu jin cheng*)' to encourage rural non-agricultural development in particular. This policy change was very significant because after it

> [r]ural non-agricultural development was now viewed not so much as a way to serve and support agriculture but as a way of generating employment opportunities so that surplus rural labour could be reallocated from cultivation to more productive activities while keeping rural–urban migration to a minimum.
>
> (Ho 1995: 361)

Four factors shown in the four case study areas

Together, as can be clearly found in my case study areas in Inner Mongolia, Shaanxi, Shanxi, and Heilongjiang, these four factors facilitated the development of TVE coalmines. It seems strange that with abundant coal and surplus labour force, these areas were poor and even suffered from coal shortages, but this was true under the former command economy system. Table 3.2 shows that in 1978 both per capita GDP and rural net income in these four areas, except Heilongjiang, were all below the national average.

While it is difficult to find data on the real rural surplus labour force both in 1978 and 1998 in these areas, Table 3.3 provides clear evidence that in these four provinces the ratio of employed persons to the total population was all well below the national level. This indicates that the surplus labour force in these four provinces might be much more than in other provinces.

Meanwhile, all these four provinces and autonomous regions were rich in coal and three of them, with the exception of Heilongjiang, were extremely rich. The

Table 3.2 Per capita GDP, rural net income and living expenditure in the four areas (yuan)

	Per capita GDP		*Per capita net income*		*Rural living expenditure*	
	1978	*1998*	*1978*	*1998*	*1978*	*1998*
National average	379	6,392	134	2,162	116	1,590
Shanxi	365	5,072	102	1,859	91	1,056
Shaanxi	291	3,834	134	1,406	134	1,181
Inner Mongolia	317	5,069	131	1,981	n.a.	1,577
Heilongjiang	564	7,530	172	2,253	125	1,465

Source: SSB 1999a.

Notes
1 At prices in 1978 (cols 2, 4 and 6) and 1998 (cols 3, 5 and 7).
2 Once again, this table applied the 1998 data instead of the updated 2002 data in order to reflect the fact that by 1998 when the government started to close TVE mines, the living standards of these rural areas were still relatively poorer than other areas in China.

Table 3.3 Population and employed persons in the four areas

	Population (1) (millions)			*Employed (2) (millions)*			*Ratio of employed (2) to total population (1) (per cent)*		
	1978	*1998*	*2002*	*1978*	*1998*	*2002*	*1978*	*1998*	*2002*
National total	963	1,248	1,285	402	700	737	42	56	57
Shanxi	24	32	33	n.a.	14	14	n.a.	44	42
Shaanxi	28	36	37	11	18	19	39	49	51
Inner Mongolia	18	23	24	7	11	10	36	45	41
Heilongjiang	31	38	38	10	17	16	32	45	42

Sources: CSA 2003: 40, 48; SSB 1999.

Table 3.4 Coal reserves and output of TVE coalmines in the four areas, 1995 (percentage of national total in brackets)

	Predicted reserve (Mt)	Reserve to be allocated to TVEs (Mt)	Number of TVE mines	Output of TVE mines (Mt)
National total	4,552,100	57,300	72,919	665
Four provinces total	1,835,600 (40)	39,900 (70)	13,425 (18)	410 (62)
Shanxi	389,900	33,300	6,700	157
Shaanxi	203,100	2,700	1,568	198
Inner Mongolia	1,225,000	3,500	2,780	30
Heilongjiang	17,600	371	2,377	25

Sources: CCC 2000: 19; Ye and Zhang 1998.

coal reserves at a depth of 2,000 metres in these four provinces alone accounted for 40 per cent of the total national coal reserves (Table 3.4).

Government policy changes propelled the development of TVE coalmines in the whole of China, but particularly in these four provinces. As the third and fourth columns in Table 3.4 show, while the coal reserve in these four areas accounted for 40 per cent of the national total, the allocation to TVE coalmines in these four areas accounted for 70 per cent of the total national reserves allocated to TVE mines. Meanwhile, although TVE coalmines in these four areas accounted for 18 per cent of the national total, their output accounted for 62 per cent of the national total, which indicates that the average coal output for TVE coalmines in these areas was higher than others. TVE mines in these areas should receive more attention because they were the most important part of the whole nation's TVE coalmines.

Great contributions of TVE coalmines

Ending China's coal shortage history

Energy shortages inhibited China's economic development and even normal household life for the decades from the 1950s to the mid-1990s. TVE coalmines' production of half of the national total ended China's coal shortage history by the mid-1990s. From 1978 to 1995 TVE coalmines produced 5.47 billion tons in total, accounting for 34.8 per cent of national output, and 25 per cent of total national primary energy. The total increased output by TVE coalmines was 3.85 billion tons, accounting for 73.5 per cent of the total national increased coal output, and 53.7 per cent of the total increase in primary energy output.

If this increased output were to have been provided by the state sector, the necessary investment is estimated to have been over a hundred billion yuan. Even though the state could have provided such huge investment, it might have had to cut substantially other necessary investment. Moreover, the state was actually not able to provide such a huge investment at the very beginning of the reform (Ye and Zhang 1998: 43). That was why the government strongly supported TVE coalmines in the 1980s and advocated that 'the state provides infrastructure, and the public invests in coalmines' (*guojia xiulu, qunzhong bankuang*).

Promoting poverty reduction, rural development and urbanization

TVE coalmines have been found extremely important in creating rural income and local revenue and in providing non-farming employment. The former Ministry of the Coal Industry (MCI) conducted the largest and most comprehensive investigation into the development of TVE coalmines in 1996, and concluded:

> The development of TVE coalmines has greatly promoted economic development and poverty reduction in the mining areas ... TVE coalmines have become the main industry of local areas driving the development of transportation, construction, chemical, power, metal and business service sectors. In poor areas with large coal reserves, most towns and villages have escaped poverty and enriched themselves by depending on the contributions of TVE coalmines. They have also made a great contribution to local infrastructure, education and social welfare.
>
> (Ye and Zhang 1998: 44–5)

As found in the case studies, TVE coalmines have changed the lives of the rural poor to a significant extent. Shenmu county in Shaanxi province, with a population of 350,000, has a land area of 7,635 square kilometres, of which 4,500 km^2 have coal reserves. In 1999 the government revenue was 180 million yuan, of which 50 per cent came from over 200 TVE coalmines. It was estimated that over 20,000 peasants were recruited into TVE coalmines, earning at least twice as much as they earned from the land. It was hard to estimate how many local residents were absorbed into coal-related business, including power plants, transporting, coking, construction, and services. Before 1986 in Shenmu there was only one small coal-fired power station, and electricity was in severely short supply. In 2000 there were sufficient power stations, and electricity was even surplus to local demand. The luxury Shenmu County Government Building does not reflect the fact that, until 1984 when the central government announced 'making the stagnant water flow fast', Shenmu was still designated a key poor county (*zhongdian pinkunxian*) and relied on the government's poverty relief funds for decades.[4] By that time (1984) coal production was around 300,000 tons, government revenue was only 10 million yuan.

It was the contribution from TVE coalmines that ended Shenmu's long history of dependence on the special subsidies from central government for a 'key poor county'. Officials in Shenmu county government and Shenmu Coal Administration still remembered that it was in 1986, and in Shenmu county, that Premier Zhao Ziyang renounced the slogan of 'to let stagnant water flow quickly' (*youshui kuailiu*). Since then TVE coalmines have developed dramatically and Shenmu's economy has been completely transformed.

TVE coalmines made a great contribution not only to government revenue, but also to rural family income and rural urbanization. Research on migration in Shanxi, based on interviews with 100 migrant households and 100 non-migrant households in 1998, showed that over 75 per cent of local rural migrants worked in TVE coalmines with an average 3,333 yuan per year remittances contributing to their family income. Those families with mining migrants not only had a much higher living standard than other families without migrants, but also had children with far greater access to various education and training opportunities (Zhang 2000). Similar findings were made in each province I visited. Moreover, those areas where TVE coalmines are concentrated have been gradually developed into towns or commercial centres from being villages or even uninhabited.

Finally, it is worth noting that the improvement of both family income and the local economic situation greatly assisted local social and human development. Shenmu is a county of Shaanxi province, but neighbouring Baode county is in Shanxi province. Historically Shenmu was much poorer than Baode so that Shenmu women sought to marry Baode men while leaving many of their brothers unable to get married. Since the development of TVE mines in Shenmu was much faster and better managed than Baode since the 1980s, Shenmu became richer while Baode was still poor.[5] For that reason, Baode women now do their best to marry Shenmu men.

Improving the structure of China's coal supply and demand

Most of China's coal reserves are in western and northern areas. Shaanxi, Shanxi, Inner Mongolia, Xinjiang and Guizhou account for 91.7 per cent of the total national predictable reserve under 2,000 metres in depth (CCC 2000: 18). However, the major coal consumption areas are on the east coast. In 1997 east China accounted for 26 per cent of national coal consumption, but only produced 15 per cent of the national total, while Shanxi, Inner Mongolia, and Shaanxi produced 34.8 per cent but consumed 14.5 per cent of the national total. Such coal reserves, production and consumption needs dictate that the basic feature of the coal demand–supply structure is 'transporting coal from north to south, from west to the east (*beimei nandiao, ximei dongyun*). This structure adds to the already severe pressure on the chronically over-strained infrastructure capacity. The average coal transportation distance increased from 426 km in 1980 to 548 km and then to 561 km in 1995 (CCC 2000:68) and average cost of transportation was over 70 yuan per ton in 1997 and over 120 yuan per ton since 2000.[6] Eventually railways became a bottleneck in the coal supply and pushed the coal price substantially higher than its production cost.

Naturally TVE coal was in high demand from local customers as it was not only easily available but also much cheaper than SOE coal. This was especially so in east China, where coal was in high demand, but most coalfields were small and scattered. As these were not suitable for extraction by large SOEs, the advantage of the TVE mines was fully utilized. From 1980 to 1995, TVE coal output in eastern China increased from 52 Mt to 237 Mt, an annual increase of 11 per cent. It was estimated that without this production, at least two more long-distance railways of the capacity of the Daqing Line (Datong to Qinghuangdao) would have been needed to transport coal from west to east.

Creating a competitive coal market in China

Numerous researchers have found that the SOEs' monopoly position was the major cause of chronic shortages under the command economy.[7] In China's coal industry, SOE coalmines had a near monopoly position, producing 85 per cent of coal by 1978. SOEs' lack of incentive and effective management to maximize their productivity, together with government's lack of investment to build new coal-producing capacity, became a major cause of coal shortage.

Thousands of TVE coalmines entered the market with great competitive advantages, which completely destroyed the SOEs' monopoly situation and changed the market structure, in which SOE and TVE coalmines each comprised half of the national output by 1996. In explaining why TVEs have comparative advantages when competing with SOEs in general, a number of factors have been highlighted by researchers, such as China's institutional structure that facilitates cooperation through implicit contracts among community members (Weizman and Xu 1994); the capability of TVEs to adapt and configure their strategy in response to the external competitive environment (Luo *et al*. 1998, 1999) and close relationship between local governments and TVEs (Jin and Qian 1998). These findings are obviously true in China's coal industry.

However, in addition to these factors, this study found more. Many TVEs were actually owned by powerful local officials or even cadres who were working in SOE coalmines. TVE coalmines also invest much less in facilities, land recovery, environmental protection, and social obligations than SOEs, and this is the key to understanding their low costs but high accident rates (see next section). Most importantly, TVEs are able to use the cheapest rural surplus labour force and so adopt a more flexible wage and welfare system.

On the surface, the advantage of TVE coal was that its costs and therefore prices were much lower than those of SOEs. Table 4.2 shows that in 1995 while the coal production costs for TVE coalmines were from 27 to 66 yuan per ton, the national average cost of SOE coal was 107 yuan per ton. In 1998 when the coal costs of TVE mines in Jixi area in Heilongjiang province were about 40 yuan per ton (including local taxes and fees), SOEs such as Jixi Mining Bureau were producing coal at a cost of 217 yuan per ton. However, the different cost structures shown in Table 4.5 reveal the underlying reason: the fundamental cause for such cost differences was that the wages and social welfare costs at TVE mines were much lower. In 1995 while wage costs of TVE coalmines ranged from 5 to 27 yuan per ton, the national average spending on wages and welfare by SOE coal bureaux was 38 yuan per ton; in 1998 while the cost of wages of the nearby TVE coalmine was 17 yuan per ton, Jixi's wage and welfare spending was 40 yuan per ton.

Making use of an unlimited rural labour force, TVE mines could not only spend much less on wages, but also adopt a completely flexible employment structure, because they were able to recruit or dismiss employees from rural areas in any number at any time, at any level of wage or welfare commitment. Also benefiting from the flexible employment structure and unlimited rural surplus labour force, TVE workers were more productive, despite their lower wages and poorer working conditions. SOE coal workers rejected the intensification of work while comparing their pay and position with their formerly higher wage levels and better social position. TVE coal workers worked as hard as possible to keep their jobs while comparing their pay with the much lower (or zero) income from the land. Therefore, the unlimited rural labour supply was the fundamental cause of greater competitiveness of TVE coalmines compared with SOEs, and even with newly established modern large coal corporations, which are still not able to dismiss employees freely and obviate the previous costly welfare system.

Consequently, with a lower price, better service and strong local government support, TVE coalmines attracted more and more customers. It was the first time that SOEs faced great competition, lost market share and suffered operating losses. Especially as coal had been in surplus supply since the middle 1990s, customers had more choice, so they had higher expectations of coal quality, service and price competitiveness. The coal surplus also enabled the government finally to liberalize coal prices in 1993 without being concerned about the possible impact. A truly competitive coal market finally emerged, much earlier than in most related industries such as electricity and railways.[8]

Opening up possibilities of building up modern firms and consolidation of SOEs

As noted, fierce competition from TVE coalmines has already forced SOEs to reform themselves. While coal supply was sufficient to provide for the whole economy and competition had been introduced into the market, the government started to neglect some of its previous responsibilities. Importantly, it decentralized the key SOEs passing them to provincial level government control, gradually ending subsidies for loss-making SOE coal bureaux and more recently even allowing the worst ones to go into bankruptcy. By passing on these responsibilities, central government made it possible to concentrate its limited funds on building globally competitive indigenous coal corporations such as the flagship Shenhua Group. The Chinese government granted the Shenhua project preferential loans of over $9.2 billion during its construction period from 1985 to 2005, making it the third largest investment project in China. This compares with a figure of $1.1 billion for China's total investment in the coal sector in 2001 (CSD 2002: 53) (see Chapters 4 and 5).

TVEs mine seams that otherwise would not be exploited

It could be argued that TVE coalmines have actually squandered a large amount of coal resources. This is true in some respects, as the next section will show. However, it is also true that TVEs have extracted substantial coal resources which would not otherwise have been extracted by the large SOE coalmines. This includes: corner coal, extremely thin coal seams, small scattered coalfields and those in inaccessible areas. In 1995, nearly 26,000 TVE coalmines, or 36 per cent of the national total of TVE mines, extracted coal from these poor-quality seams, totalling 156 Mt, accounting for 27 per cent of total TVE coal output (Ye and Zhang 1998: 46). TVE coalmines, acting as a type of coal 'hoover' gathering coal from the least accessible areas, provide powerful evidence for the continuing rationale for these mines. In addition to absorbing surplus labour force, making use of more coal resources was another reason why closing all TVE coalmines would be uneconomic. Obviously China needs to use less investment to produce coal and also needs to make use of any coal resources by whatever methods because of its energy scarce economy. Chapter 6 will develop this argument showing how government should make use of all the advantages from different types of coal producers.

Significant problems of TVE coalmines

The great contribution of TVE coalmines should not hide their equally significant negative effects on many aspects as shown below.

Diseconomies of small scale and excessive competition

China is the largest coal producer in the world, with an output of 1.36 billion tons at the peak year of 1996. Of this, almost half was produced by tens of thousands of small TVEs. The diseconomies of scale can be found in each step from production to the final market.

Table 3.5 shows how small the scale of TVE coalmines was in 1995, which worsened the diseconomies of scale at the national level, as seen in Table 5.3. The free entry of TVE coalmines resulted in the surplus market, excessive competition, and depressed prices.[9] These developed to the extent that many SOE coal producers, and even some TVEs, could not cover their costs.

As coal production has been in surplus since 1994 and the government did not launch a policy of limiting total output until 1998, the excessive competition among tens of thousands of coalmines depressed coal prices to an uneconomic level, which simply could not cover the costs of SOE coal producers, and even of some TVEs. In 1999 the average selling price in China's coal market was 144 yuan per ton, while the average production cost for large SOE coalmines was 104 yuan per

Table 3.5 Size and structure of China's TVE coal producers, 1995

	Number of TVE producers	Total output of coal (MT)	Average output per producer (tons)	Share of total TVE producers (%)	Share of total TVE output (%)
Total TVE coalmines	72,919	579	7,943	100	100
> 30,000 tons per annum	6,646	245	36,929	9	42
10,000–30,000 tons per annum	15,164	195	12,875	21	34
< 10,000 tons per annum	51,109	139	2,710	70	24

Source: Ye and Zhang, 1998: 62.

Note
1 Total output of TVE coal in this table is based on one particular survey, and is slightly different from the data shown in Table 2.2.

ton, and the average transportation cost was around 100 yuan per ton. As TVE coal was more competitive at its much lower price, the only way that coal producers, especially SOEs, could sell more was to reduce their prices.[10] This was a very important factor behind the losses of SOE coalmines. In addition, when coal still could not be sold at a reduced price, most SOEs had to suspend production. As a result, advanced equipment and techniques in the larger SOE mines had 50 Mt of unused capacity each year, while backward TVE coalmines were still entering the market. For example, Shenhua Group possesses the best equipment and quality of coal in China, but in 2000 less than half its 11 coalmines could produce at full scale, and others had to suspend production.

Waste of large coal resources

Excessive competition, together with the lack of well-planned coalfield extraction processes and the unreasonable tax system, have resulted in large wastage. Table 3.6 shows that both resource tax and royalty payments in China's coal sector are calculated entirely on a given company's turnover. They do not significantly vary between different resource categories, resource quantity and quality, or the ease with which the coal can be extracted and transported. This, together with the lack of minimum recovery rate requirements from the resource administration body, commonly results in a 'plundering style' of extraction.[11]

TVE coalmines, motivated to save costs by cutting down on geological testing, surveys or preparatory work, usually extract coal where SOEs are developing.[12] Some of them even use the existing facilities and material of SOE mines. Naturally TVE coalmines select the best source while readily abandoning the second-best. While the recovery rate at key SOEs is as high as 50–70 per cent, it is just 30–40 per cent at the local SOEs and 10–20 per cent at local non-SOEs. This compares with 55 per cent in Britain in 1936, and 97 per cent in Germany in 1999 (Yan *et al.* 2000: 28–33; Thomson 1996: 726–50). Because of the low current recovery rate, China is therefore losing more than 4 billion tons of coal reserves each year (SCA 1999a: 11).

Environmental damage

China is the second largest sulphur dioxide producer in the world, just behind the USA. This is largely related to the fact that in China more than 76 per cent of primary energy comes from coal, more than 80 per cent of electricity is generated by coal-fired power stations, and more than 70 per cent of coal is for direct combustion. In 1998, out of a total 1,232 Mt, 310 million tons were washed. The wash rate was 25 per cent; within this, the wash rate for TVE coal was only 7 per cent (Table 3.7). Although, by 1998, China had 1,527 coal-washing facilties with the capacity to wash 494 Mt, most TVEs did not have a coal-washing facility, although they produced about 50 per cent of the total coal production. In addition to sulphur dioxide emission, a huge number of TVE coalmines extract coal in very shallow mines, which may release more ash directly into the air because the

Table 3.6 Tax categories, rates and collectors for coal since the 1994 tax reform

Category of tax[1]	Tax basis	Tax rate	Tax collector[5]
Value-added tax	Turnover	13%[2]	CG
Income tax	Income	33%[3]	CG
Resources tax	–	0.3–5 yuan/t[4]	LG
Land value-added tax	Value-added on land	30–60%	CG and LG
Royalty	Turnover × recovery rate	1%	CG 50%; LG 50%
City maintenance and construction tax	Turnover and income	5–7%	LG
Education tax	Turnover and income	3%	LG
Property tax	Rent income	12%	LG
Urban land use tax	–	–	LG
Cultivated land occupation	–	1–10 yuan/acre	LG
Stamp duty	Contract value	0.03–0.05%	LG
Transportation access tax	–	60–320 yuan / year/a company	LG

Sources: CERC 1995: 99–101; interview sources from Datong in Shanxi province, May 2000.

Notes
1 This table excludes fees charged for coal producers. Fees are charged at various rates in different areas. These fees were charged in Shanxi Province in 2000: administration fee 1% <mult> turnover; mining managing fee 1% <mult> turnover; land rent fees 1.5–2%; water resource subsidy fee and green maintain fee 1%; others 4–6%. In addition, for exporting coal out of Shanxi, except key SOE coal, it has to pay 39.8 yuan/ton to Shanxi Coal-Transporting-Sale Corporation.
2 Except for coal and a few others, the VAT is 17%. Before 1994 there was 3% 'product tax' on coal.
3 Before 1994, different income tax rates for different ownership enterprises: 55% for large and medium SOEs; 10–55% for collective-owned enterprises (COEs) and small SOEs; 35% for privately owned enterprises; 30% for foreign investment enterprises and 15% for special economic zones (SEZs).
4 1.6 yuan/ton in Shanxi.
5 CG: collection by central government; LG: collection by local government.

Table 3.7 Rates of coal washing for different types of mines in China in 1997

Type of coalmine	Coal production (1) (Mt)	Washed coal (2) (Mt)	Ratio of (2) to (1) (%)
Key SOE coalmines	526	246	47
Province/county mines	226	52	23
TVE mines	573	40	7
Total	1,325	338	26

Sources: *China Coal*, Vol. 26 No 1. (Jan 2000), as quoted in Horii and Gu, 2001: 47.

degree of carbonization of shallow coal seams, formed at a later stage, is low. Many TVE coalmines also extract coal without planning consent, and this may damage the coal seam or cultivated land. In north-east China, TVE coalmines extended their underground mines to residential areas and could even open their pitheads in houses. In north-west China, rivers have run dry because of coal extraction from the riverbed.

High accident rate

At the same time as relatively safe SOEs are unable to produce at full capacity, thousands of dangerous TVEs produce nearly half of the national coal output. Most TVE facilities are officially described as 'extremely backward' and their mines likened to 'mole tunnels'. It is hard to imagine the working conditions of most TVE mines. The most typical in north-east and north-west China are small well-style mines, whose major vertical shaft is about one metre square. In such narrow mine shafts people cannot move around easily, but dozens of workers have to extract coal and lift it out of the shaft mainly by manpower. Once the main shaft is established, the workers dig as many horizontal sub-shafts as possible radiating from the main vertical shaft to extract more coal. While SOE mines use pit props to prevent shaft collapse, TVE mines, in order to save costs, usually do not. Most importantly, this sort of small mine has only one exit, which means that if any accidents occur it is almost impossible for workers underground to escape.

Table 3.8 shows the seriousness of TVE coalmines' safety record, which was much worse than that of SOE coalmines, let alone than in any other major coal-producing countries. Government authorities admit that the rate of 'severe and particularly large accidents were consistently high', and this rate had worsened in the very recent past. Official data show that since 2000 the numbers of deaths have been consistently over 5,600 per year, making death rate per million tons as high as 4.5 on average, and the death rate in TVE coalmines was as high as over 10 per million tons (SCA 2000, 2001, 2002b, 2003). From January to April 2003 there were 611 accidents with 1,090 deaths. Although the overall death rate per

Table 3.8 Relative safety of different types of coal producer in China 1999–2003 (deaths per million tons)

Coal producers	1999	2001	2002	2003 (Jan–April)
China				
of which:	4.500	5.311	4.641	3.580
Key SOEs	0.966	1.487	1.270	1.010
Local SOEs	3.458	4.682	3.743	1.970
TVEs	10.990	18.501	12.171	13.380

Sources: SCA 2003, quoted in SDRC 2003: 9; SCA 2002d; SCA, 2000; Nolan 1999: 63.

Note
By 1995 the death rate per million tons was 0.04 in the UK, 0.25 in Poland, and 0.96 in Russia. Since the mid-1990s in advanced coal-producing countries such as the USA and Australia, the death rate per million tons has been almost zero.

million tons dropped, the death rate of TVE coalmines actually increased. Although overall accidents dropped by 6.5 per cent, the large accident rate (with deaths of between 10 and 29) increased by 12.5 per cent, of which many happened in TVE coalmines. Unofficial estimates were even higher than these official numbers, because it is common knowledge that TVE mines usually tend not to publish details of accidents and deaths.

There has been special concern about legal or illegal TVE coalmines that extract coal from the same coalfields as SOEs, which would greatly interrupt the SOE's normal production where the TVE mines have damaged the coal seam. Sometimes their shafts were even connected, so that many accidents initiated by TVEs (such as flooding) caused severe damage to the SOE property and production.

Moreover, many small mines do nothing to control dust. The Chinese government acknowledged that thousands of miners die of lung diseases every year in addition to those killed in accidents.

The difficulties of closing TVE coalmines and the underlying causes

Major pressures to close TVE coalmines – challenges from transition and globalization

A policy on TVE coalmines 'closing coalmines and decreasing output' (*guanjin yachan*) was announced in November 1998. Since then, TVE coalmine development has turned to decline, after their meteoric rise. The basic cause of the closure of TVE coalmines, according to the government, was because of the detrimental effects of these mines, as shown in the last section. However, research has shown that there are deeper causes behind this closure policy. A document issued by the State Council showed that the government was well aware of the negative problems of TVE coalmines as early as in 1990, stating:

> Compared with large and medium enterprises, their [small enterprises'] technology is backward, their consumption of materials and energy is huge, and their product quality is poor ... *They compete with large and medium [SOE] enterprises for materials, energy, capital, and markets, making large and medium enterprises unable to fully utilize their production and technical potentials. From a social point of view, this trend of downsizing and diversification will inevitably aggravate the inefficiency in industrial production.*
>
> (State Council 1990, quoted in Lin and Yao 1999: 13, emphasis added)

This quotation indicates that what concerned the State Council was not just the general negative consequences resulting from TVE production, but also the more particular negative impacts on the large and medium-sized SOE coalmines. It is hard to explain, if this assumption is incorrect, why the State Council – which was well aware of the TVE mines' problems a decade ago – did not force them to close earlier. In contrast, between 1992 and 1998 TVE coalmines developed even more dramatically. The basic reason was that in 1990 SOE coalmines were not experiencing the same financial difficulties as they are now and the challenges from globalization were still distant. The challenges of development, furthermore, were then still comparatively great as coal and energy remained in short supply, while the rural sector was still very poor. However, by 1998 SOE mines had grown to become the principal loss-makers among all industries, with a loss of 3.7 billion yuan, (compared with the loss of 1.9 billion yuan by the textile industry, the second biggest loss-maker).

SOE mines would be unable to survive faced with the TVE mines' double impact: their plundering style of extraction and their huge scale of production. As regards the production style, TVE coalmines usually extract coal at the same coalfield as SOE companies, posing a severe accident risk for both TVE and SOE mines. Such accidents usually entail much heavier property losses for SOE than for TVE mines. At the Working Conference on the National Township- and Village-Owned Mines held in March 1994, the vice premier Zou Jiahua expressed great concern about the many problems of these mines. He made it very clear that 'the state would not ignore the plundering that threatened the coal seam in the key state-owned coalfields and that TVE mines would be rectified and straightened out' (Horii 2001: 53). With regard to the huge production from TVE mines, it resulted in severe market surplus. TVEs' intense price competitiveness and more effective management caused SOE coalmines to become uncompetitive and then huge loss-makers.

However, as shown in our case studies on SOE coalmines, the transition of SOEs from their original status in the command economy system incurs enormous challenges in many ways. It poses a challenge to the government's finances, to the current social security system, to SOEs' recognized social status, to the available institutional structure, and to social stability (see Chapter 4 on challenges brought by transition). Facing these challenges, the government could not allow many SOE mines to go bankrupt or close, even though that was the inevitable consequence. The only possible way to save SOEs in the short run was to control the excessive market competition. Now that the decades of coal shortage had ended, it was time for the government to review the TVE problem and turn the policy from encouraging their rise to enforcing their demise.

Another underlying cause of TVE coalmines being closed was the concern that globalization will pose serious challenges for the current diseconomies of scale in China's coal industry. As early as 1990, as shown in last quotation, the State Council (1990) expressed the concern that the boom in TVE coalmines had resulted in the 'trend of fragmentation and diversification' and this (trend) would 'inevitably aggravate the inefficiency in industrial production'. By 1996 policy-makers had realized that new growth strategy should focus on better achieving economies of scale and technical progress rather than only on investment of more resources and labour (SPC 1996: 6). The building of Shenhua Group is based precisely on such considerations. Besides, China has expressed its ambition in its Tenth Five-Year Plan for the Coal Industry to build seven more cross-sector and cross-regional large coal corporations by restructuring the whole industry. Naturally, worries about failure increase with China's progress towards accession to the World Trade Organization and all the major global mining companies entering the Chinese market (see Chapter 5). From this perspective, therefore, it is not difficult to understand why the government had to make the decision to close tens of thousands of small and technically backward TVE coalmines.

It therefore can be concluded that difficulties in dealing with the challenges from transition and globalization were the fundamental causes for the government's closure of problematic and uneconomic TVE mines. These hidden intentions were clearly reflected in the process of implementing the closure policy. On the one hand the government announced in 1998 that three kinds of TVE coalmines were subject to the closure policy: the illegal ones (lacking two necessary licences), those producing low-quality coal with a poor safety record, and those extracting coal in the same coalfield as large SOE coalmines and threatening the safety of the SOE mines. This indicates that China still needs the better-quality TVE coalmines and the focus of closure was to solve problems facing TVEs. On the other hand, during the implementation of the closure policy, quotas giving the number of TVE mines to be closed and the amount of output to be reduced were also issued to local government. This policy reveals that the central government's major purpose was to solve the problem of surplus supply and to improve the economy of scale.

Difficulties of closure

It was becoming more and more necessary to close TVE coalmines, as we have demonstrated above. However, the central government could not make the decision until 1998. The closure of TVE mines since 1998 has also been proved extremely difficult, especially during the period between 1998 and 2000. In 2000, even after two years of closure policy, it was found that in Shanxi the coalmines were reopened after closure; in Inner Mongolia new licences were still being issued by local government in spite of central government prohibition; and in Heilongjiang, where working groups responsible for closure threatened to blow up privately owned mines, owners went as far as staying in their mines and declaring: 'if you want to blow up my mine, blow me up first'.

There are doubts about the government statistics for the number of TVE mines closed. In 2000 it was announced that the proposed goal for 1999 to close 25,800 TVE coalmines (from a previous 72,000) and to reduce output by 250 Mt had been achieved (SCA 2000). In early 2002 the State Economic and Trade Commission (SETC) announced that, after implementing the closure policy for three years, remaining TVE coalmines numbered only around 20,000 in 2001 (Shi 2002). The latest report from the same source shows that by 2002 there were 23,500 TVE coalmines. However, in 2002 when the closure policy was slightly loosened in response to high demand, TVE coal output immediately soared from 264 Mt in 2001 to 418 Mt (see Table 2.2). Many believe that such a dramatic increase would have been impossible if the TVE coalmines had in fact been reduced to a third of earlier numbers.[13] Local officials who were in charge of the policy implementation admitted that many of the closures were actually of mines already or nearly out of production, or completely illegal coalmines lacking all licences. Some TVE coalmines were even responding to the closure policy by 'working during nighttime, but sleeping during daytime'.

The fundamental cause of such strong resistance to closure was still poverty and the unstoppable desire to eradicate that poverty. It is not surprising that local governments are desperate to protect TVE coalmines, since they form an important local revenue and employment source. In addition, as demonstrated, developing TVE coalmines in most areas was an initial impulse in order to create a network of local industry and business, and this even had a positive effect on the local social services such as schools and hospitals. Once coal production stopped, all related businesses would run down. This would be a disaster for many local governments.

The closure of TVE mines means much greater losses for their owners. It is well known that TVE coalmines were strongly encouraged to develop rapidly at an earlier time. In addition, most TVE coalmines were established using private funding. Even in Shaanxi, where coalfield conditions are much better than in most other areas, the investment required to open a TVE coalmine with a production capacity of 500,000 tons per year ranged from one to five million yuan in the 1980s, and this figure excludes a large amount of invisible 'investment' which might also be needed to bribe the different authorities and pay various unreasonable fees. If the closed TVE mines were legally set up, they had every reason to be compensated, at least by retrieving their initial investment.

For workers in TVE mines, the closure meant a huge loss, even more important than their lives. We have noted the terrible death record in TVE coalmines. It is not because these workers did not care about their lives – they simply had no alternative. As found again and again in TVE mine accidents, most of the miners killed come from poor rural areas with high unemployment rates, willing to jeopardize their lives to earn a living. Survivors expressed themselves more directly: 'This mine is dangerous. But where can I earn this 200 yuan a month if I leave here?'; and 'without money, why should I need a life?' (SCA 2002c). Table 3.2 shows that by 1998 when the closure policy was implemented, these coal-rich provinces were still poorer than the national average, though much better than they were in 1978. This situation was stayed the same until 2002.

From this we see that resistance to closure was the direct result of a fear of loss of income. If so, it seemed unwise for the government not to compensate for these losses in order to reduce resistance. It was understood that, whether from the viewpoint of justice, or from that of making the closure easier, compensation was a necessity. However, since 1998 when the closure policy was implemented, TVE coalmines were forced to close without, in most cases, the topic of compensation being raised. In a few cases in Shaanxi province, independent auditing agencies were invited by the provincial closure working group to estimate the losses resulting from TVE mines being closed, but none of the TVE mines which were closed had received any compensation by the time of my interviews.

The only case of TVE mines being successfully closed by compensation rather than by administrative force was that of Shenhua, which powerfully proves the importance of compensation to the success of closure. In this case, Shenhua leased the right to extract coal from the TVE mine at a fee of 8–10 yuan per ton until all the coal had been extracted from the leased mine (also see Chapter 5). Local government has been extremely supportive of the lease. On the one hand, many local TVE coalmines either already suffered financial losses or loss of confidence because of the surplus coal market and the government's closure policy. On the other hand, leasing a mine to Shenhua could mean receiving even more income by doing nothing. In the case of Shenhua leasing Yujialiang mine as a blending coal base, it brought over 40 million yuan of revenue to Yujialiang's owner – Dianta Township in Shenmu county, Shaanxi province.[14] Compared to the previous five million yuan annual government revenue, it was easy to understand how the county viewed Shenhua as a great 'saviour'. In return for Shenhua's contribution, Dianta Township government closed all eight TVE mines which were extracting coal in Shenhua's coalfield.

This case indicates that local governments still have sufficient authority to implement the central government's closure policy. *If they are willing to do so, they are able to, but the prerequisite was that they would receive more benefits than that from contribution from TVE coalmines.* However, this revives the old question: whether the government is able to provide the huge amounts of compensation. As will be seen in Chapter 4, while the government could not afford the huge transition costs including the clearing-up of bankrupt SOEs, it would be equally difficult to provide the compensation to closed TVEs.

The Chinese government indeed can be criticized for policy failures on TVE coalmines, such as the failure to have a long-term policy with regard to development. This will be examined in Chapter 6. However, the difficulties of avoiding such failure are also obvious. If China's coal production had not turned into surplus after the mid-1990s, if TVE development had not threatened SOE survival and China's capacity to compete with global giants, the government might never have been determined to close them. Here one can see how the challenges from transition and globalization forced the government to close TVE coalmines. However, as closing TVE coalmines would certainly be counter to the desire of hundreds of thousands of rural poor for development, the resistance to the closure would be enormous. The conflict between the forces of development, transition and globalization is therefore also fully reflected in the cases for and against the closure of TVE coalmines.

Conclusion

Development is a supreme challenge for China. The underlying force behind the dramatic development of TVE coalmines was the rural population's desperate need to eliminate poverty. Making use of a cheap and unlimited rural labour force, TVE coalmines became more competitive and profitable. In return, they made an enormous contribution to rural development by providing over three million jobs and much higher family income and local revenue, and to the whole nation's economic development by providing almost half the nation's coal output. These achievements confirm the effectiveness of the development strategy presented by Lewis (1954) for underdeveloped states with a huge surplus rural labour force. First, development in such a situation heavily relies on transferring the surplus rural labour force from farming to non-farming sectors. Second, surplus labour can be used instead of capital in the creation of new industrial investment projects, or it can be channelled into nascent industries, which are labour intensive in their early stages. Third, government should play an important role to initiate and encourage such transformation.

Meanwhile, TVE coalmines' dramatic development has created a competitive market environment, which drove the SOE coalmines to make serious reforms. Both TVEs' newly established productivity and SOEs' improved productivity provided sufficient coal, becoming a fresh impulse for the general economic development. China's improved economic performance in turn reduced the resistance to transition. From this point of view, the development of TVEs has also created an innovative transition model for China, which assisted the smooth and effective transition from the former command economy.

It should not be denied that TVEs have had huge negative effects on the economy and environment. However, once again, when poverty is still widespread and development is a supreme challenge, TVEs are still in demand despite their negative effects. While TVE coalmines have been continually criticized for their poor safety record and resource wastage, people should ask why their workers are willing to do such dangerous jobs to earn such a modest wage, compared to other sectors' workers and other classes in the society. When a country is still afflicted by severe poverty, to earn a living might be judged more important than pursuing human rights or future 'sustainable development'.

However, the decline of TVE mines' is inevitable in the longer term because China has to confront the challenges from transition and globalization as well, which are mainly reflected in the necessities to restructure the industry, to cope with large loss-making SOE coalmines and to build globally competitive modern coal corporations. As the government's financial capacity and current institutions are not well placed to transfer SOEs effectively and quickly, the government has to press for the closure of TVE mines to solve the surplus market so that SOEs can be kept functioning. As economies of scale and global competitive capacity must be pursued to deal with challenges from globalization, small, unsafe, and uneconomic TVE coalmines must be closed in the restructuring of the whole industry.

As closure of TVE coalmines will certainly work against the desire to eliminate poverty for hundreds of thousands of rural poor, the resistance to the closure is enormous. The conflict between the three challenges from development, transition and globalization is therefore clearly reflected in the struggle between pro- and anti-closure sides. The development of the TVEs, and even the development of the country as a whole, depends on how the government will deal with these inter-related challenges in a well-balanced and compromising way. While either closure or non-closure of these mines will endanger the balance of the government's economic policy, restructuring them might restore that balance while at the same time accelerating mergers among them, clarifying their ownership (for example by setting up joint-stock companies – see Sun 2002), regulating their resource, production, and taxation and by presenting a clear signal for their future development. This will be discussed in more detail in Chapter 6.

Figure 3.4 TVE coalmines in China

Top left: A TVE coalmine in the popular well style in Heilongjiang in north-east China. This is the most dangerous sort of mine as it has only one exit which means if any accidents happen it is almost impossible for underground workers to escape. *Top right:* a TVE coalmine in Shanxi province, central China. This coalmine even produces coke to export to Germany, but pollution of the famous Yungang Grottoes nearby goes unmonitored. *Bottom left:* improved but still very poor rural countryside – Baode county in Shanxi province. *Bottom right:* a TVE coalmine in Inner Mongolia, north-west China. The smoke in the photos is from the coal's self-combustion. It indicates that in these remote high mountains where coal is so richly in reserve, but waSted, and people wish to climb out of poverty, TVE coalmines are at their most advantageous.

4 Transition

Transforming traditional SOE coalmines

Supporters of strategy A ... emphasized the creation of favourable conditions for bottom-up development of the private sector ... In contrast, strategy B's emphasis was on the rapid elimination of state ownership ... Ten years past, the practice has proved that strategy A was superior to strategy B.

(Kornai 2000: 1–2)

I want to ask you [Russian people] for forgiveness, because many of our dreams have not been realized, because what we thought would be easy turned out to be painfully difficult. I ask for forgiveness for not fulfilling some hopes of those people who believed that we would be able to jump from the grey, stagnation, totalitarian past into a bright, rich and civilized future in one go. I myself believed in this. It seemed that with one spurt we would overcome everything ... In some respects I was too naïve. Some of the problems were too complex.

(B.N. Yeltsin, President of Russia, 1991–9, Resignation Statement, 31 December 1999, quoted in Ellman 2000a: 1417)

Introduction

The quotation from Kornai above is probably the best explanation for the difference between the two post-communist reform strategies. By this definition, China's reform so far has followed a pattern of strategy A ('gradualism'), while Russia's reform might be the best illustration of strategy B ('shock therapy'). There is an increasing realization that China's gradualist reform strategy has outperformed the shock therapy strategy (see Kornai 2000). However, there has always been a strong case against the gradualist strategy and this is normally based on the large losses and redundancies incurred by state-owned enterprises (SOEs) (see Åslund 2002).

In this view, the performance and efficiency of SOEs have lagged far behind non-state enterprises, and the avoidance of privatization in China on the grounds of preserving social stability may be 'overlooking the social tensions being created by the asset stripping, corruption, and macroeconomic instability caused by the unreformed ownership structure of the SOEs' (IMF 2000: 121). In this interpretation, gradualism is not a contributor but a constraint on China's success (e.g.

World Bank 1996; Sachs and Woo 1997; IMF 2000; Pack 2000). China's SOE reform policy since the 15th Party Congress held in 1997 to 'grasp the large, release the small' (*zhuada fangxiao*), was taken as evidence that 'China has not been an exception to absorbing the positive international experience with the privatization of SOEs' (Sachs and Woo 1997: 29). Since China joined the World Trade Organization (WTO), SOEs have faced even fiercer pressures to downsize, privatize, or even close as soon as possible. In 2003 the newly created State Asset Management Commission (SAMC) is widely assumed to be the signal for a new wave of exits by SOEs. A rapid decline of SOEs is generally predicted (Chinaed 2003).

These arguments raise several fundamental questions: how to understand China's gradualism and its consequences, especially compared with the difficulties faced by SOEs. If gradualism is a source of China's economic success, how is it possible to explain the huge financial difficulties that SOE coalmines are facing? If gradualism is a constraint on economic success, is it possible to explain how China's economy grew so quickly and so smoothly? Was the gradualist approach wrongly applied in reforming SOEs in the past? Could the chronic SOE problem have been solved through more dramatic action? Could it be the case now? Should China give up its gradualism as it has joined the WTO? This chapter attempts to answer these questions.

As large loss-making SOEs are always considered to be the major evidence for the weakness or failure of China's gradualist reform, this chapter uses a case study of a typical traditional state-owned enterprise in financial difficulties, the Jixi Mining Bureau, to demonstrate the following argument.[1] Gradualism was more favourable and feasible for China during earlier reforms because of the stronger challenges from transition, and it is still vital today because of changing business practices throughout the world and because of still complicated issues which underlie SOE problems. This may shed light on the prospects for SOEs in China.

The argument is analysed in four sections. The first section presents the typical gradualist reforms in China's coal industry. The second section documents the extreme financial and operational difficulties from which SOE suffer. The third section analyses whether the difficulties have arisen from gradualist reforms. The fourth section assesses whether dramatic privatization was more favourable and feasible during earlier reforms and is still vital today. A conclusion is then reached.

Gradualist reforms implemented in the coal industry

As demonstrated in Chapter 2, the coal industry has been facing a more urgent but also more difficult challenge from its transition because of its irreplaceable role in China's industrialization, because of having to catch up with progress in the rest of the world, and because of the consequences of being pervaded by the command economy system. Since the reforms of 1978, especially, coal was becoming more crucial for the economy, and improving the productivity and efficiency of the industry became one of the central tasks.

While orthodox transition economics recommended privatizing large and medium-sized SOEs to promote their transformation, China deployed a sequence of typically gradualist reforms, to which the coal industry was no exception. These reforms not only fundamentally turned the coal market from severe shortage to surplus in order to support faster economic growth, but also dramatically changed the sector's ownership structure: from SOE domination before 1978 to one in which SOE and TVE coalmines each shared half of the national output by the mid-1990s.

Supporting the development of TVE coalmines

The first gradualist reform in China's coal industry was to support the fast development of non-state-owned coal producers – mostly TVE coalmines – while keeping SOE coalmines unprivatized. As shown in more detail in Chapter 3, the pressures from the severe coal shortage during the 1980s forced the government to encourage the development of TVEs beginning in the mid-1980s. The first direct consequence was that 73,000 TVE coalmines had sprung up by 1998 because of government encouragement. The second consequence was that China's coal market turned from severe shortage to surplus. The third consequence, most relevant to this chapter, was that, for the first time, SOE coalmines faced a competitive challenge and consequently incurred huge losses, because TVE coalmines could undercut their costs and prices.

Implementing the dual-track coal price system

The second gradualist reform was first to introduce a dual-track price system and gradually liberalize coal prices until 1993, rather than completely liberalize prices immediately after 1978. Depressed coal prices during the decades from 1949 to 1993 was an important cause of the losses incurred by SOEs. Figure 4.1 shows that from 1985 to 2000, except for 1994, the price of Jixi's coal was continually lower than its cost. This data is consistent with that of all key SOEs.

After long delays and negotiation, the dual-track price system was finally adopted in the coal sector in 1985. The new system took the planned output of 1984 as a baseline quota. Coal production below this quota would be sold at the planned price; that which was above the 1984 quota but below that for 1985 could be sold at 50 per cent more than the planned price, the guide price; that above the 1985 quota could be sold at a price 100 per cent higher than the planned price if it was used for government plan purposes, otherwise it could be sold at a free market price (Ge 1991: 200). Although the dual-track system has incurred much criticism, it actually stimulated coal output and improved coal producers' profits at the beginning. For loss-making SOEs, it also helped to subsidize part of the loss. For example, from 1985 to 1989 key SOE coalmines made a total of 7.333 billion yuan gain from price increases (Ge 1991: 201).

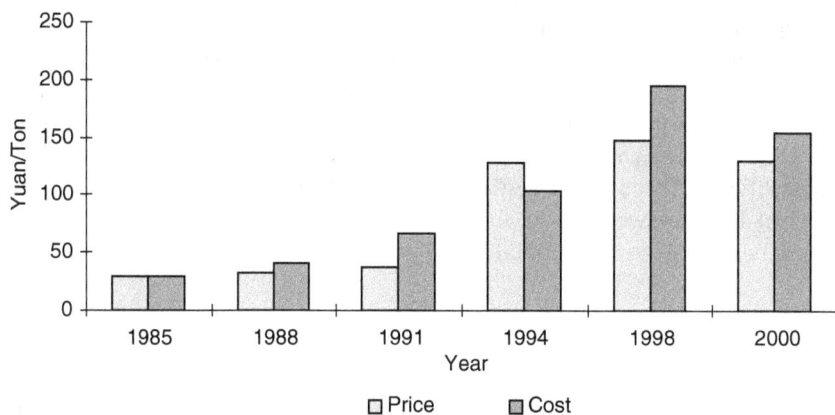

Figure 4.1 The cost and price of coal at Jixi Mining Bureau from 1985–2000

Sources: Jixi Mining Bureau Statistic Bulletin data, 1985, 1988, 1991, 1994, 1998, 2000.

Implementing the 'contract responsibility system'

The third gradualist reform was to implement a 'contract responsibility system' (CRS), to focus on improving the competitive capacity of SOEs through decentralizing more management autonomy to firms rather than by privatizing them. The first CRS implementation period for key SOE coal bureaux ran from 1985 to 1990. Three targets were set for the period. The first was an output target. All key SOE coal bureaux were asked to increase their output by an average 20 Mt each year, and reach 500 Mt by 1990. The second was an infrastructure investment target, which was a total of 31.5 billion yuan for all key SOE bureaux. The third was a financial target, which was maximum losses of 300 million yuan for all the key SOE bureaux and coal machinery factories. Each SOE coal bureau was required to reach a financial target, i.e. there was a maximum loss it could make. The CRS reform was expected to bring a number of benefits for SOE coalmines, including the fact that they could make more profits as they were allowed to go 50 per cent over the planned price for the coal within the contracted target, and 100 per cent over the planned price for the coal which exceeded the contracted target. Meanwhile, SOEs could retain the part of the profits made because of this policy change if they did not exceed their financial target (CCEC 1999: 456–7).

Separating government functions from direct involvement in economic affairs

The fourth gradualist reform was to improve the competitive capacity of SOE coalmines by separating government functions from direct state involvement in economic affairs. From 1983 the central government took the previous year's subsidy as a base, and set a subsidy target to the key SOE coal bureaux, which was

300 million yuan each year afterwards. The government expected the subsidy to be kept within this ceiling during the Seventh Five-year Plan (1986–90) and to be discontinued thereafter, in the hope that SOE coalmines would have benefited from the other gradualist reforms introduced at the same time, such as the increased coal price following the dual-track price reform, and better management following the CRS reform. When coal supply was sufficient for the whole economy and competition had been introduced into the market after the mid-1990s, the Ministry of Coal Industry was finally abolished in 1998, and all key SOE coalmines belonging to it were transferred to local provincial government.

Enforcing hard budget constraints on SOE coal bureaux

The fifth gradualist reform was to discipline SOE coalmines by enforcing strict budget constraints on them. The most important changes linked to administrative reform were in the handling of investment. Before the 1980s, all of SOEs' fixed capital investment and over 50 per cent of working capital investment came from the central government budget. At that time the liability ratio of SOE coalmines was quite low.[2] Since 1985, fixed capital investment has been moved from budget allocation to a bank loan system and all working capital investment must be administered through the banking system. Since no 'capital fund' (*zibenjin*) has been injected, SOE coalmines' liability ratio has been growing larger and larger.[3] This is especially true for newly built and reconstructed coalmines, which are under more pressure to repay loans and interest to the bank. In 1993 the State Development Bank (SDB) of China was established as a state policy bank responsible for lending to all SOE coalmines. The bank issues loans based on its own detailed investigations and evaluations, which has forced SOE coalmines to adopt strict budget constraints.

Reforming the tax system on coal

The sixth gradualist reform was of taxes. Under the command economy system, so as to stay in line with the unreasonably low controlled price for coal, the tax on coal was also kept as low as possible, at only three per cent of revenue. However, since the 1994 tax reform, the tax on coal has consisted of a 13 per cent value-added tax, which is four per cent lower than that on other commodities, but 10 per cent higher than previously. Fearing the potential impact on coal producers of such large tax increases, the government adopted a strategy of first collecting tax on coal, and then later refunding part of the tax revenue to some enterprises as an open subsidy.

Establishing a new social security system

The final gradualist reform is reflected in the evolutionary progress from the formal socialist welfare system, characterized by the life-long employment, free accommodation and free medical services provided for SOE workers by their companies, to a modern social security system which is more compatible with a market economy system. The first stage for workers who were made redundant is to be sent to a 're-employment centre' at their enterprise, and they receive a basic living allowance linked to local living standards. Two years later, in the second stage, they are entitled, as unemployed, to receive benefit from local government, which is then supposed to take over responsibility from central government. After a further three years, at the final stage, the unemployed have to rely on themselves if they still cannot find jobs. At the same time the social welfare responsibilities of firms will also be transferred to the local community. As a condition for local government to accept responsibility for redundant people, central government, firms and employees of the firms must each make contributions in advance, in principle each contributing one third.

The difficult situation of SOE coalmines

Before examining whether the above gradualist reforms resulted in SOEs' financial difficulties, it is necessary to examine how serious the condition of these SOEs had become.

It should not be denied that most large SOE coalmines have experienced extreme financial difficulty and a high level of worker redundancy. With the exception of 1997, the coal industry as a whole had been the largest loss-maker among China's industrial sectors from 1984 to 2000. Losses amounted to 3.7 billion yuan in 1999, and 2.7 billion yuan in 2000 (see Table 4.1). With the implementation of the 'closing mines and depressing output' (*guanjin yachan*) policy since 1998, the industry was reported to have made a profit of 1.5 billion yuan in 2001 and 2.3 billion yuan in 2002, much of this being the profit contribution of 25 key coal enterprises in 2001 and 32 key coal enterprises in 2002 (in other words, the other SOEs – around 3,000 in total – made virtually no profits) (SETC 2002, 2003).

However, a large proportion of wages, pensions and redundancy payments are still waiting to be paid. Triangular debts[4] (*sanjiaozhai*) has been a consistent problem for most SOE coal companies, and was as high as 30 billion yuan by the end of 2000, 55 per cent of that year's coal industry's revenue. There were 20 key SOE coal companies needing to pay living allowances of 100 million yuan to those made redundant, 17 companies needing to pay 284 million yuan in pensions, and 66 companies needing to pay 7.3 billion yuan in wages. Although these companies might show a profit in their accounts, they lacked the cashflow to operate their business normally. Meanwhile, more than two million (about 25 per cent of the total) mining workers had lost their jobs since 1994, when the large-scale redundancies started.

Jixi Mining Bureau (hereafter Jixi) is an appropriate case to illustrate and assess these problems because it has experienced a dramatic change, from being a showcase SOE coal bureau to being one of the most troubled. Jixi Mining Bureau is located in the city Jixi in Heilongjiang province, a province in the far north-east. Of the 104 former large SOE mining bureaux, it is the second oldest and the only one to have existed during the rule of the Qing Dynasty, Russia, Japan, the

Table 4.1 Profit/losses in various industrial sectors 1997-2000 (hundred million yuan)

	Coal	Textiles	Construction materials	Petro-chemicals	Machinery	Non-ferrous metals	Electronics	Medicine	Electric power	Gold	Tobacco
1997	12	-45	-14	166	99	-4	145	70	82	7	126
1998	-37	-19	-16	110	86	-14	160	78	95	10	117
1999	-37	10	34	309	153	0.4	200	98	83	11	101
2000*	-33	49	63	850	152	50	311	106	67	9	103

Sources: *People's Daily*, 8 January 2001; SCA 2001.

Note
* Data for 2000 is for January to October only. The total loss for 2000 was 2.7 billion yuan.

Republic of China and the current People's Republic of China (*wuyi qizhu*). It used to be the second largest mining bureau, and its output reached 20 Mt in 1991, second only to the Datong Mining Bureau. Jixi has produced over 600 Mt of coal in total since it was taken over by the Chinese Communist Party (CCP) in 1948. Moreover, due to its far north-east location, it is also the only mining bureau which has made a large contribution to the wars in which China attacked Japan, supported Korea and defended the former Soviet Union, not only through its coal for energy, but also through its miners acting as temporary soldiers. Naturally, people in Jixi used to be very proud of their history and identity.

Under the former command economy, Jixi enjoyed government guarantees for finance, secure input supplies and output sales. The workers also enjoyed the highest wage level among all industrial sectors. Even during the 1980s, when a market economy system had already been introduced, Jixi still did not experience any difficulties. Before 1986 Jixi made a loss of 20 to 30 million yuan each year, and the ratio of liabilities to assets was only nine per cent. An important background condition in explaining this was that coal was in severe shortage until the 1990s. Demand for coal was so high that the bureau did not have to be concerned about being unable to sell enough coal. Rather, it enjoyed significant benefits from those consumers who urgently needed coal.

However, since the mid-1990s Jixi's situation has grown steadily worse, falling to its lowest level in 1998 (Figure 4.2), when the output was reduced to 6.53 Mt, and losses reached one billion yuan, some one quarter of the sector's total loss. The difficulties by 2000 were reflected in the following aspects.

Severe financial difficulties

Jixi's total assets were four billion yuan, but its total liabilities were at least 4.33 billion yuan by January 2000. The ratio of liabilities to assets was 98 per cent, the ratio of short-term debt to cashflow was 152 per cent. Its bank credit rating was reduced to BB, which meant Jixi no longer qualified for loans. Liabilities included

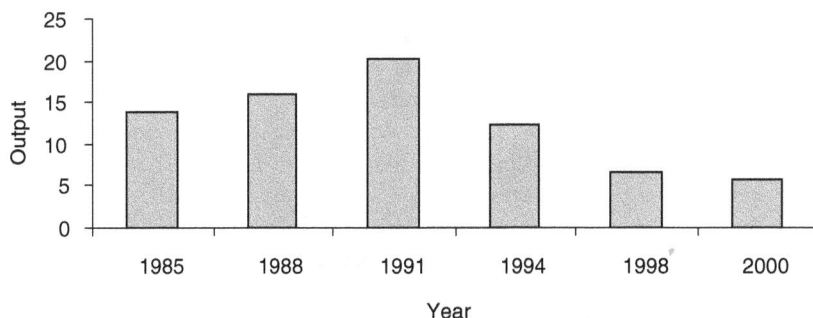

Figure 4.2 Coal output in various years for Jixi Mining Bureau (Mt)

Sources: Jixi Mining Bureau Statistical Bulletin, various years.

the postponed payments of 430 million yuan in salaries, and 81 million yuan in pensions.

Chaotic operation

The high level of liabilities and lack of cash made Jixi's operation chaotic. For example, if Jixi's debts to electricity suppliers were not paid, the electricity supply would be stopped; if its debts to rail companies were not paid, these companies would refuse to transport Jixi coal. Funding shortages led to a chronic under-investment in the exploration of coal faces, in technology improvements, and in safety and security. This has resulted in severe shortage of coal faces and consequent decline in output, and also pushed the death rate up to as high as 2.88 deaths per million tons (compared with 0.96 in other SOEs and zero in Shenhua). It is, therefore, not surprising that the largest accident in China's coal industry since the 1950s occurred at Jixi in June 2002, when 129 were killed.[5]

Losing customers

Existing customers were lost, and payments from customers was only 41.3 per cent of total debts (compared to Shenhua's 100 per cent). A large part of payments from customers was not in cash but in kind – the customer's own products, cars, rice or even wine from Russia – a practice known as *mo zhang*.[6]

Brain-drain

The problem of a 'brain-drain' further upset and depressed people in Jixi. On the one hand there were large losses of highly qualified engineers and trained employees, while on the other new graduates were not attracted. This was sharply different from the situation before the 1990s when new graduates strove to find employment in Jixi.

Partial bankruptcy

Finally, four coalmines out of Jixi's total of 13 were bankrupt in 1999, which increased total redundancies to 33,654 at the same time the number of active employees were reduced to 51,000 and pensioners were increased to 52,576 (see Figure 4.3).[7] By 1998, workers and pensioners did not receive wages or pensions for as long as 36 months and they were still waiting for more than 20 months' salary and pensions by 2000. Meanwhile, redundant workers had 218 yuan (about $24) per month as a 'basic living allowance' to provide for their whole family which on average consisted of at least three people.[8]

There were several adverse social consequences. First, life for redundant workers was extremely hard. Second, the high incidence of broken families increased. Third, crime rates increased and people lost their sense of security. There were numerous examples of family members of redundant coal workers committing

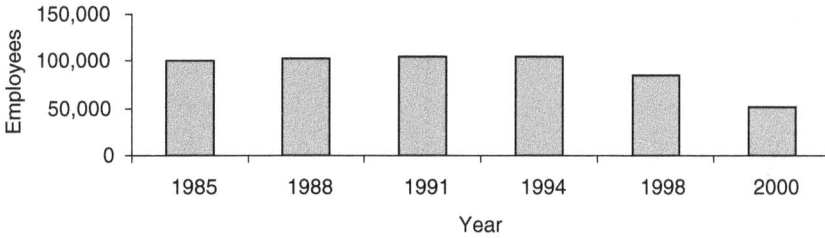

Figure 4.3 Number of employees in Jixi Mining Bureau

Source: Jixi Mining Bureau Statistical Bulletin, various years.

suicide or being forced into crime and prostitution. Fourth, workers were disillusioned and angry with the corruption of local mining cadres and with the unreasonable policies of central government. Consequently, demonstrations frequently took place and social instability became a threat. Since the Didao coalmine staged its first demonstration in 1992, there have been more than 400 demonstrations by January 2000, with actions including stopping railways for 11 days, and demonstrations at the Heilongjiang provincial government building and the former Ministry of Coal Industry at Beijing.[9]

Three categories of the causes of Jixi's difficulties

It is necessary to identify all the factors that contributed to Jixi's difficulties before examining whether, and how, these factors were related to the gradualist reforms listed in the first section. The data collected from the Jixi case study by 2001 suggests three major categories of causes of Jixi's difficulties. The first category included those basically natural causes beyond the control of Jixi Bureau: natural events, historical legacies, and the external economic environment. The second category comprises causes resulting from competition from TVEs and modern coal groups. The third category includes man-made causes, in particular those arising from government policies and Jixi's internal management.

The first category of causes

The first category comprises causes from natural, geological, historical and economic environment factors. Natural conditions must be considered first. The Jixi coalfield has been developed for almost a century, and its main shaft is now more than 1,000 metres deep. Compared with the 20 to 30 metres depth of Shenhua's coalmines and the shallow TVE coalmines, the production cost of Jixi is naturally higher due to the extra spending on material, electricity, and manpower.[10] Table 4.2 shows that Jixi's materials costs were 68 yuan per ton in 1998, compared with 5 yuan per ton for Shenhua Group in 1999. Jixi's total cost was as high as 216.90 yuan per ton in 1998, compared with 66.48 yuan per ton for Shenhua Group in 1999. More crucially, despite higher expenditure on materials,

Table 4.2 Cost structure in China's four major categories of coal producers in 1995, 1998 and 1999 (yuan/ton)

	Key SOEs average 1995[2]	Open-cast mines 1995	A TVE mine in Inner Mongolia 1995	A TVE mine in Heilongjiang 1995	A TVE mine in Heilongjiang 1998	Jixi Mining Bureau 1998	Shenhua Group 1999
Unit cost of raw coal	107	34–63	27	64–66	40	216	67
Of which:							
Materials[1]	23	8–24	–	8–9	5	68	5
Wages[1]	38	5–12	5	27	17	40	3
Electricity	9	5–9	1	4	2	32	6
Depreciation	11	5–9	–	–	–	27	6
Repairs	6	3–4	–	–	–	32	5
Maintenance	7	–	–	–	–	6	7
Other	14	5–24	8	2–3	2	11	33
Local taxes/fees	–	–	13	23	14	–	2

Sources: Nolan 1999: 63; Jixi Mining Bureau Statistical Bulletin 1998: 32; Shenhua Group 2000; SCA 2000.

Notes
1 Including welfare funds.
2 The average cost for SOE coal companies in 1999 was 103.8 yuan/ton.
.

electricity and manpower, Jixi's output was lower than Shenhua's. While Jixi employed 53,000 workers to produce 5 Mt of coal in 2000, the single Daliuta Mine of Shenhua group employed 350 employees to produce 8 Mt of coal. The unit cost of Jixi was, obviously, much higher than that of Shenhua.

With regard to historical origins in this category, the first cause was that imposed by the command economy system. 'Between 1949 and 1985 the international [coal] price rose 400–600 per cent, while in China the average price rose only 239 per cent' (Thomson 2003: 127). This led to the fact that not only Jixi, but also the whole coal sector made hardly any profit (Figures 4.4 and 4.5). Figure 4.4 shows that Jixi did not make a profit after 1985, and actually never made a profit in its history. Figure 4.5 shows that the entire coal industry made virtually no profit before 1985

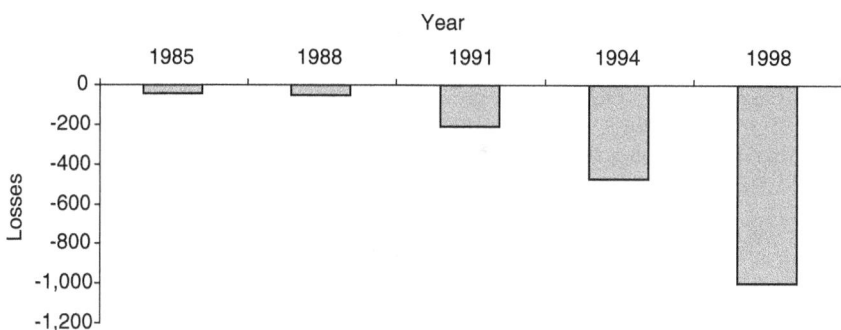

Figure 4.4 Losses in Jixi Mining Bureau (million yuan)

Sources: Jixi Mining Bureau Statistical Bulletin, various years.

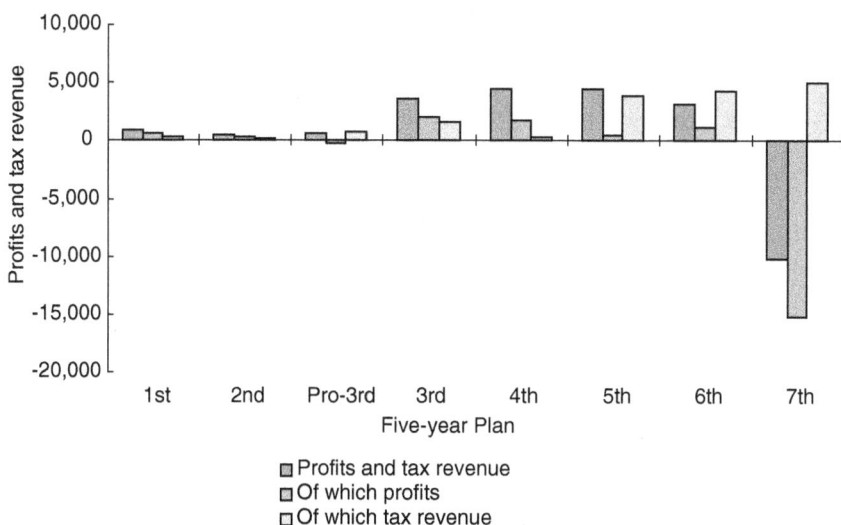

Figure 4.5 The balance sheet for key SOE coalmines 1953–90 (million yuan)

Source: CCEC 1999: 560.

(the end of the Sixth Five-year Plan) and became substantially loss-making after the Seventh Five-year Plan (1986–90). Loss-making appears to have grown quickly after implementation of a series of gradualist reforms for transition.

The second historical origin in this category was that, due to Jixi's long history, there was an accumulation of expenditure on social obligations, pensions and welfare.[11] The expenditure on these social obligations and welfare was 160 million yuan in 1999, which added 40 yuan per ton to the cost of coal. The average ratio of active workers to pensioners for China's SOE coalmines as a whole was 3.6 to 1.0, but in Jixi it was 1.4 to 1 before the four coalmines went bankrupt. In addition, being the only large SOE in Jixi City, Jixi Mining Bureau was supposed to share many other social obligations with the Jixi City government. For example, there were 172,000 households in Jixi City using electricity, of which Jixi Bureau supplied 100,000. Although the cost of producing the electricity to Jixi Bureau was 0.42 yuan/kwh, the price at which it was sold to Jixi households was limited to only 0.36 yuan/kwh. This single obligation required Jixi to find an additional 70 to 80 million yuan each year. It may seem surprising that Jixi Bureau did not raise this unprofitable price, but even the Bureau itself believed that there was no justification to do so. It had supplied low-price electricity to Jixi City since the start of the command economy, and Jixi City's residents have taken it for granted, making Jixi Mining Bureau hard to overturn such a firm assumption.

In terms of its wider environment, the basic problem was that the social and economic background was not favourable for Jixi. North-east China is a well-known base for heavy industry and military firms. Some of them have been customers of Jixi's coal for decades. However, after the reform, China turned its attention away from heavy industries and, after the cold war, the government decided to restructure and downsize most military firms. When these firms were closed or had severe financial difficulties, the demand for Jixi's coal fell and the debts owed to Jixi were not paid.

The second category of causes

The second category of causes arose from intense competition between TVEs and large coal groups. We have seen in Chapter 3 that TVE coalmines developed dramatically and became major rivals of SOE coalmines. We have also seen that one of the causes of the uncompetitive capacity of SOEs was that the cost of SOE coal was much higher than that of TVEs, including much higher wages and welfare for their employees, and higher investment and expenditure on facilities, land recovery and environment protection (see the 'wage' and 'maintenance' columns of Table 4.2). While SOE companies like Jixi lacked both the incentives (mainly due to the ownership problems) and the means (mainly due to lack of funding) to reduce its coal cost, or to bribe customers, TVE mines had a stronger incentive to bribe railway managers and consumers to transport or sell more coal, because TVE mines had relatively clearer ownership and normally belonged to a limited number of people. As a result, by 2000, almost all Jixi's former consumers in the power station sector had moved to TVE coalmines.

Another factor was that the TVE coalmines were strongly supported by local governments because of their contribution to local government revenues and family incomes, and because they absorbed rural surplus labour (see Chapter 3). Naturally, when any conflict emerged between SOEs and TVEs, local government supported the TVEs.

It is well known, but not officially recognized, that most TVEs were actually owned by powerful local officials or cadres working in SOE coalmines. When customers came to Jixi, even Jixi's cadres themselves induced customers to buy coal not from Jixi Bureau, but from TVEs, in which they shared interests or actually owned. One TVE coalmine, which shared the same coalfield as the Xiaohengshan coalmine of the Jixi Bureau, occupied the best-quality part of the coalfield. It was contracted out from Xiaohengshan coalmine. The mine manager used to be a cadre from Xiaohenshan, the equipment could be borrowed from Xiaohengshan and workers came from those made redundant from Xiaohengshan, as it was already bankrupt.

Moreover, in Jixi, corrupt cadres leased more and more of Jixi's railway facilities to TVEs, which enabled TVE coal to be sold to remote customers and to compete with Jixi across a much wider area. Currently in China many TVEs are still limited in their profits due to being unable to access railway transportation.[12] In Shenhua, because Shenhua controls its own railway, TVEs have to sell coal to Shenhua rather than compete with Shenhua directly. However, this was not the case in Jixi. Jixi created and strengthened its own competitors by contracting out their best coalfield and renting out their railheads one by one. The obvious reason was that cadres could benefit from such leases and contracts.

In addition to obvious competition from TVE coalmines, there was a strong unseen threat of indirect competition from large modern coal groups such as Shenhua. On the one hand, modern coal groups can outcompete traditional SOEs through their use of advanced technology and equipment, and consequently lower production cost (see Table 4.2). On the other hand, government concentrated most investment, which was previously scattered across traditional SOEs, on these large groups to build globally competitive coal corporations. For example, when Shenhua Group was awarded $9.2 billion of investment (see Chapter 5), investment in traditional SOEs almost entirely ceased. To some extent, the threat of cessation of investment was much more important than losing some market share in determining Jixi's fate. When coal was already in surplus and the government's funds were limited, the cessation of investment indicated that it was not necessary to save the large loss-makers.

The third category of causes

The third category was of causes resulting from government policies and Jixi's internal management. With regard to the government policies, three major changes on price, subsidies and internal management of firms were most closely related to Jixi's difficulties. The dual-track price system actually stimulated coal output and improved coal producers' profits, and, from 1985 to the early 1990s, helped loss-making SOEs to reduce their losses. As Thomson (1996: 740) records, the large loss-making situation,

> had actually begun to improve in 1992, thanks to the government's decision ... to allow a larger and larger share of total coal production to be sold at market price. In that year the industry's losses fell by 500 million yuan and in the first half of 1993 they fell by another 420 million yuan, making a total loss reduction rate of 27.8 per cent.

However, coal prices rose much more slowly than other prices. For example, by 1990, the output proportion of major industrial materials whose prices were still set by government had been reduced to 44 per cent, but it was still as high as 95 per cent in the coal industry. From 1985 to 1989, the price of industrial materials was up by 68 per cent, but SOEs coal price was up by only 23 per cent, and the coal market price was up by 41 per cent (Pu 1992: 5). This means that while most industrial materials adjusted their price to a relatively reasonable level, coal prices were actually further away from their real value than before. Jixi's cost table (Figure 4.6) clearly reflected that most of their cost increase from 1985 to 1991 was due to material and electricity price increase. The cost of coal in Jixi in 1985 was 59.84 yuan per ton in 1985 and 93.69 yuan per ton in 1991, up 57 per cent. Among the cost components, materials were up 8 per cent, and of electricity up 5 per cent. The cost component with the biggest increase was 'miscellaneous payments' ('other payments' in Figure 4.6), up 16 per cent. However, detailed data show that the largest elements in 'miscellaneous payments' were on small repairs, office expenditure and management fees, which was still mostly due to the price increase of various materials and services. Jixi's management concluded: 'we added 0.10

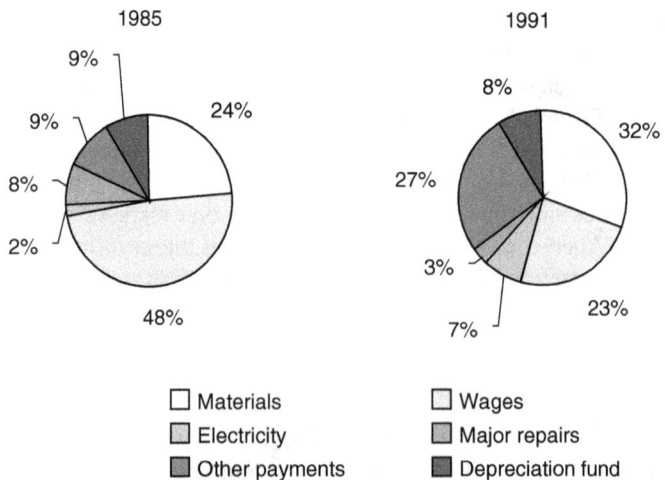

Figure 4.6 Comparison of the cost structure of Jixi coal in 1985 and in 1991

Sources: Jixi Mining Bureau Statistical Bulletin 1985, 1991.

yuan per ton to the coal price, but the gain was diminished completely by the increase of 0.20 yuan/kwh for electricity' (Ren, Z, interviewed at the Jixi Mining Bureau Conference Hall 2000).

Moreover, coal prices were liberalized too late to generate enough profits to cover SOE losses. When the government expected SOE mines to reduce their losses through a large increase in the price of coal after the final liberalization of the coal price in 1994, there had already been in fact about five years of coal surplus.[13] When TVEs' coal was so cheap and abundant,[14] raising the price would only make SOE coal more difficult to sell and would drive more consumers to TVE coal. This was the key reason why losses by SOE mines grew dramatically after the mid-1990s, with losses in one year being greater than five years' losses in the 1980s (see Figures 4.1 and 4.5).

The gradual withdrawal of government subsidies was directly responsible for losses by SOE mines. Coal industries in few countries apart from Australia, South Africa and the USA make large profits. For this reason most governments subsidize their coal industries through grants, price support, limitations on coal imports or supporting long-term agreements between producers and large consumers. Figure 4.5 shows a large deficit in China's coal industry. During the 38 years from 1953 to 1990, there were 13 years of industry loss-making. As stated before, the government expected to reduce and finally cease the subsidy in the hope of letting the coal industry benefit from dual-track price reform and the CRS reform. However, economic reform in China became more dramatic from the mid-1980s, and several changes meant that actual loss-making was far greater than the target set by the government. Among these changes, the prices of most products were increased dramatically, much faster than that of coal; the wages element of coal production cost rose from 6.16 yuan per ton in 1980 to 17.4 yuan per ton in 1990, an average rate of 10.9 per cent each year; the tax rate on coal was increased from 3 per cent to 13 per cent;[15] and finally the investment policy was changed from the state budget distribution system by government to the bank loan system, which increased repayments on loans and interests for SOEs.[16]

Alongside the policy reform on price and subsidy, the CRS reform was implemented in order to create a market and to stimulate competition through decentralizing more management autonomy to firms. This has been widely recognized as a contributor to China's successful economic reform as a whole. It also contributed to Jixi's great increase in output until 1992. However, the CRS reform was a major cause of Jixi's decay.

From 1986 to 1993 Jixi signed an eight-year contract with its governing body – the former Ministry of Coal Industry (MCI). Although Jixi was not able to stay within its loss-making target,[17] the Jixi managers at that time did not like to disappoint the then leader of the MCI. So on the one hand they kept reporting that Jixi's losses had been greatly reduced. On the other hand, they endeavoured to get funds from other sources to ensure that the contracted output target could be achieved. Moreover, with increased autonomy under the contract responsibility system, 'managers can do whatever they want to do' (Hassard *et al.* 1999: 63). Because there were so many expenses that managers needed to take into account (including their own benefits) with limited funds, naturally, the first consequence was a shortage of funds.

The second, related problem was short-termism by management. The then managers had to meet output targets, but investment in building new capacity and improving working conditions was severely restricted. One possible option for management to ensure that output targets were met was to freeze investment in R&D, safety, and exploration. These short-sighted methods can be illustrated as follows: workers extracted coal which should not have been extracted at that time. As people in Jixi said, 'we are eating the food of our grandsons'. More workers were asked to extract coal while fewer were developing coal faces for further extraction.[18] Cadres focused on output while sacrificing investment in facilities, equipment, new technology and even safety and security. The bankruptcy of the Xiaohenshan Coal Mine of Jixi was precisely the consequence of this sort of short-sighted view. The Xiaohenshan Mine used to be a model of modern coalmines. However, since coal was extracted too intensively and related works were neglected, by 2000 there were no coal faces prepared for further extraction, no funding to develop new coal faces, and not even any updated maps for the development of new coal faces. Although Xiaohenshan still had 100 Mt of coal reserves, it was made bankrupt in 2000.

In this third category, Jixi's internal management was found to be one of the major sources of Jixi's difficulties. In addition to the short-term management described above, the director at the time was said not only to be short of management experience but also extremely corrupt. To stay in favour with the Ministry of Coal Industry, the director focused heavily on short-term output while neglecting the development of new coal faces; he misrepresented Jixi's position by false accounting; offered posts based on personal preference and cash payments from candidates; transferred state owned coalmines to private hands to make profits; and even awarded portions of the Jixi coalfield as a favour to friends. This director was the leading cadre of Jixi from 1992 to 1998, during which time coal output dropped from 20 Mt to 6 Mt and the losses increased from a very low level to one billion yuan, making him the key figure presiding over Jixi's decline. While people in Jixi wanted him to return for punishment, he was, in contrast, promoted to be the deputy director of Heilongjiang Coal Administration.

The success of gradualism versus SOE's difficulties

The costs of gradualism

The difficulties of traditional SOEs, such as Jixi, had by no means only one single cause, as has been shown in the last section. The most obvious difficulties were rooted in China's former command economy system, and from even earlier than that. The evidence is that, even without gradualist reform, Jixi as well as most of the traditional SOE coalmines, rarely made any profits throughout their history, mainly due to low coal prices under the command economy, and also because of the specific nature of the coal industry. Coal industries throughout the world have rarely been profitable, except in the USA, Australia and South Africa. Naturally, with time and with deeper mines, coal production cost will grow. This is one reason why Germany, Britain and other countries with a long coal-mining history put huge subsidies into their coal industries.

Gradualism was, however, a contributor to SOE difficulties. Examining the gradualist reforms in the coal industry as shown in this section, they more or less contributed to Jixi's difficulties. Allowing newly established TVE coalmines to compete with non-privatized Jixi mines destroyed Jixi's monopoly rent and forced it into an unfamiliar market. However, facing market competition, non-privatized Jixi could not reduce its costs because of its welfare obligations and because of the lack of incentives to improve its productivity. Meanwhile, liberalizing SOE coal prices too late and too slowly in comparison with TVE coal prices and prices for other materials brought Jixi to an even worse position. The CRS reform allowed SOEs to retain their profits after fulfilling output targets, infrastructure investment and subsidy, which certainly encouraged short-termism among Jixi's managers.

There was also clear evidence to show in the case of Jixi that ambiguous ownership was one of the causes of those difficulties. This was reflected in: (1) Jixi's lack both of incentives and the means to reduce its coal production cost and to attract and retain customers; (2) state assets, such as high-quality coal reserves, could be transferred to the private sector without any transparent exchange or even payment; and (3) managers could still be promoted in spite of incompetence and corruption. These problems arose because SOEs are state-owned, so there is either no personal sense of responsibility, or top-level decision-makers are ignorant of the situation at grass-root level. Those implementing management decisions are responsible for following orders and ensuring achievement of targets during their period of responsibility, rather than the interests of the firm in the long term.

As a key SOE, Jixi's one director and nine deputy directors were all appointed by the Department of Organization of the Chinese Communist Party, Heilongjiang Provincial Branch. The director had no authority to control the deputy directors. This caused lack of cooperation, and also was the cause of cadres being more responsible to their political masters than to their colleagues or those below them. Even if their performance was poor, managers could still be promoted if higher-level officials regarded them favourably. From this point of view SOE reform in China is not just an economic, but also a political task.

The contributions of gradualism

However, despite the cost, gradualism contributed most especially with regard to the short-fall economy at the start of reform. While all former socialist countries faced the same problems of encountering shortages during transition, China focused on economic development by adopting a gradualist transition to serve this purpose. It made use of production from both the recently established TVEs and the non-privatized but restructured SOEs.

During the early reform period, when coal shortages severely constrained economic development and SOE coalmines provided over 85 per cent of China's coal, protecting what was virtually the only available and reliable energy source was definitely a positive force for economic development. Without gradualism, TVE coalmines could not have been developed so quickly, and coal shortages would have remained unsolved. The average annual GDP growth rate of nine per cent might then have been impossible, with local development and rural industrialization lagging behind. In Russia's case, SOEs were privatized before resolving market shortages, leading to a sharp downturn in living standards.

Because of the problems resulting from the gradualist reforms in China's coal industry, such as the dual-track price system, there were indeed mistakes concerning the sequence and degree of price liberalization among different sectors which disadvantaged the coal sector. However, there were corresponding advantages for many other sectors in which coal was a major input, such as steel, transportation, power generation and light manufacturing industries. These sectors had either reduced production costs, or increased profits or exports, and consequently the whole economy benefited from the sacrifice made by the coal sector. In 1990, state industry had losses of 35 billion yuan, of which 34 per cent were losses first in the 'coalmines' and second in 'oil and gas'. This was caused mainly by the low planned prices (Wang 1998: 56). According to many economists (Chang and Singh 1993; Wang 1998; Nolan and Wang 1999), this sort of loss-making is not always inimical to economic development.[19]

Most importantly, gradualism ensured that China's whole economy would remain stable and allowed the government to adjust prices and remedy mistakes one by one, thus avoiding serious dislocations due to complete price liberalization, which is also associated with irreversible reform. Shock therapy is less attractive if government focuses on the whole nation's development, rather than on just one product or sector.

When comparing the achievements of China with those of Russia, there is no reason to claim that the costs of deploying gradualism in China were greater than those resulting from attempting to follow a faster path to a market economy in Russia.

Challenges in transition in favour of gradualism

It must be made clear that although rapid privatization of SOEs looked in principle much more desirable than gradualist reforms, as it was expected to solve all the SOEs' problems, such as price distortion and heavy-burdened welfare obligations at one go, it was not feasible in practice, at least not in China's coal industry. Although no large SOEs had been privatized in China's coal industry, the bankruptcy of large SOE coal companies provides a clear indication of the lack of feasibility of their privatization. In 1999, 14 SOE coal bureaux comprising 40 coalmines were forced into bankruptcy (SCA 1999b: 3), of which 4 of Jixi's 13 coalmines were forced into bankruptcy by the central government.

The proposed bankruptcy was much more complicated to resolve than expected. The first critical problem met by the 'Bankruptcy Working Group' was that they could not find suitable articles in China's only relevant Law on Bankruptcy (*pochan fa*) to fit Jixi's case. The objective of the law was the legal entity, which in this case was Jixi Mining Bureau rather than its four proposed bankrupt coalmines. In addition, there was no previous bankruptcy case in China's coal industry as a precedent. Consequently, the group had to use an 'innovative way' to deal with this new situation.

The second problem was that no one was willing or able to afford the huge bankruptcy costs, including the writing-off of the banks' bad debts, one-off payments for redundancy, and payments to local government for social security benefits for redundant workers. Without the central government's up-front payment at the time of bankruptcy, local government would not admit Jixi's redundant workers into its local insurance system (*difang xitong baoxian*) from their previous sector insurance system (*hangye xitong baoxian*) during the three years in which the redundancies occurred. There were three proposed sources for paying the bankruptcy costs: the government's new fund support; annual subsidies; and cash from selling the assets of the four mines.[20] However, it proved very difficult to sell the assets because no-one wanted to buy assets of a declining industry in a remote part of China, consequently their price was greatly reduced.

According to the group's estimation, the total cash raised by asset sales would be able to pay 15 per cent of the total cost of bankruptcy. This meant the remaining 85 per cent would have to come from central government, but central government also lacked the financial resources to meet all these costs. Therefore, Jixi Mining Bureau negotiated with both central and local government for better treatment. This resulted in prolonged negotiations taking place between the three parties.

The third problem was that many of Jixi's liabilities had been transferred to the balance sheet of the coalmines that were about to become bankrupt. This was connected to the fourth problem, namely the absence of a neutral and fair auditing and legal process. This made it possible to falsify company accounts.

The final problem was that the bankruptcy increased the number of redundant workers to 33,654 compared with only 51,000 remaining active workers. The group admitted this was its most serious problem because coal cities had evolved as more coal was discovered, so that there were few jobs outside the industry for those made redundant from it. This resulted in suffering for a large number of families, substantial migration of redundant workers to other regions, and hundreds of strikes and demonstrations.

All these problems were challenges. Besides, transition is also a major challenge for the tradition and prestige that SOE coalworkers had enjoyed from the 1950s to 1980s. Jixi coalworkers were proud of their history and identity: they were 'the most glorious class'; they received the highest level of wages among all industrial sectors; they had fought against Russia when Russia tried to occupy Zhenbaodao Island; they used to produce over 20 Mt of coal a year and were ranked as the second largest of the SOE coal bureaux in China; they maximized their output whenever economic conditions demanded; graduates sought jobs in Jixi before the 1980s and women found coalworkers attractive marriage propositions. They were still fully supported by the government through subsidies until the late 1980s because their coal was still in high demand. Only after the coal market turned to surplus in the 1990s, and the government ceased most investment while withdrawing some subsidies, did Jixi start experiencing difficulties. All the interviewees admitted that they found it extremely difficult to accept Jixi's dramatic fall, although, to some extent, they understood it was unavoidable.

Due to these challenges, the government of China had to repeatedly delay the proposed bankruptcy plan. By June 2000 only 53 out of 565 SOE coalmines were bankrupt. None of them was able to be privatized simply because no one was able to deal with so many related problems, such as the removal of social obligations. If bankruptcy, such as that at Jixi and elsewhere, is so difficult, why should one expect privatization to be easy? Aware of the possible difficulties, China started to implement privatization of small SOEs as late as 1997, twenty years after the initial launch of reform and when basic living standards were secure. Privatization was generally perceived as extremely fast and successful (Lau 1999). However, as large SOEs represent the major concentrations of employment, capital and assets, and most of them are in essential industries, they are still mainly controlled by the government.

Lessons from other transitional economies

The disastrous consequences brought about by the rapid privatization in Russia and other transition economies indicate that China's gradualism was a wise choice. Rapid privatization in Russia, as Kornai admits, did not produce clearer property rights in a 'people's capitalism'. By contrast, the 'fair distribution' of property was 'short-lived'; 'ownership was quickly concentrated in the hands of a few. The sales of state assets (at fair prices) did not redistribute wealth or income, nor did it increase the wealth of the state', but was 'coupled with mass, manipulated transfers of property to managers and privileged bureaucrats' (Kornai 2000: 2–3). Moreover, these new owners, due to lack of confidence in their own country, stripped or transferred state assets overseas rather than redeploying and restructuring them to promote the national economy. Among the consequences were 'widespread impoverishment, a decline in employment, growth in unemployment, increased inequality, a decline in public-provided services, social exclusion and in some countries, a worsening of the health of the population' (Ellman 2000b: 139). As stated at the beginning of this chapter, Yelstin, the sincere practitioner of the shock therapy, had to ask the Russian people's forgiveness for his naïve assumption that all the old problems could be solved in one go.

Besides Russia, other East European transition experiences also caused Kolodko (1998: 23) to remind the outside world that: 'one should not rely on misguided analogies with experience from distorted market economies. One must consider the specific features of that type of emerging market', and 'if institutional arrangements are left to the spontaneous process and unleashed forces of the liberalized market, corruption and organized crime will occur in extreme cases'.

The deep causes of the failure of the shock therapy reforms in so many transitional economies are by no means only a problem of personal naivety, for which Yeltsin had apologized. Rather, Stiglitz (1999a) argues that the failures go deeper, to a 'misunderstanding of the foundations of a market economy as well as a misunderstanding of the basics of an institutional reform process'. Stiglitz analyses several major problems in developing countries, which lead to privatization being nearly impossible or only making the situation worse:

> Restructuring theories eventually became more nuanced. It was recognized that foreign buyers did not want to be saddled with the politically sensitive tasks of cleaning up for past mistakes. It was recognized that investors wanted a 'pure play' consisting of the assets that fit into their business plans, not the whole bundle of assets that seem to have been randomly agglomerated in the past. It was recognized that the very uncertainty created by the mixed bag of assets and liabilities decreased the price that buyers were willing to pay. ... It was recognized that the government had to clean up the soft loan and environmental problems on its own and could not expect foreign buyers to come in and bail everyone out. Now governments are taking more responsibility for what is called 'passive' restructuring to clean up the balance sheets and reduce the labour problems (e.g., with active labour market policies and improved social safety nets) – all still without the 'active' restructuring decisions about product lines and new machinery that should be left to the buyers.
>
> (Stiglitz 1999b: 60)

Indeed, transition, especially a dramatic transition, poses great challenges for a nation's available financial capacity, financial institutions, and for the quality and legitimacy of its administration, even where democratic and long-established. There was no reason for not being cautious when implementing transition in such an immense country as China.

The success of gradualism reflected in SOEs' difficulties

Although the difficulties faced by SOEs are always an emotional issue, SOE decline should be seen as a positive effect of China's gradualist reforms. Jixi's case makes it clear that SOE difficulties were not inherent from the beginning of the reform. Rather, they started after the expansion of TVEs and because of intense competition with them. Until the early 1990s Jixi still enjoyed an 'easy life' with no concern for the market because coal was in shortage, no concern for funding because the government continued its huge investment, and no concern for job losses because more workers were needed in order to produce more coal. The situation of SOEs was indeed better than at present, but the whole nation's economy was suffering from shortages, which were closely related to the monopoly problem. At the present stage, however, mainly since the 1990s in the coal sector, the entry of TVEs has destroyed the monopoly position of SOEs, creating intense competition and turning the market from shortage to surplus. It is since then that the SOEs have progressively deteriorated. However, the nation's economy has significantly improved. Such dramatic changes came precisely from gradualism. Taking the nation's economic growth as measurement, gradualism must be judged successful.

Moreover, it is the success of gradualism that made deepening the reform of SOEs possible. After the TVE coalmines had delivered sufficient output to meet the high demand for coal, and the coal market had even turned to surplus, the government lost its fear that the restructuring of SOE coalmines might result in disruption of production and economic growth. In the late 1990s, therefore, it started to sell off most small SOEs while restructuring large ones. Moreover, after two decades of gradualist reform, in which they had failed to compete with TVEs, SOEs more and more acknowledged their own fundamental weaknesses, and realized that their disappearance was inevitable. This means that gradualism has also ensured there was sufficient time for people to accept the reform so as to reduce their resistance to it. Moreover, two decades of development in China's economy have created more new jobs for those made redundant, and even TVE coalmines are able to absorb some of the unemployed from SOEs. While both gradualism and shock therapy force SOEs to reform, gradualism reduced the suffering of SOE employees and reduced the political threat.

Finally, the success of gradualism also made the building of globally competitive large coal corporations possible. As shown in Chapter 4, while lack of economies of scale and scope have made China's coal industry suffer excessive competition and high transaction costs domestically, the world coal industry has fundamentally changed to oligopoly. The urgent task for China's coal industry policy should be to build indigenous competitive coal corporations to achieve economies of scale so as to compete in the global market. Shenhua is a successful example of this.

However, to build more coal corporations like Shenhua, the government has to enforce the entry controls on TVE coalmines, to close unproductive SOE coalmines, to establish new social welfare provision, and to concentrate its limited capital to support a reasonable number of large groups. The key point remains that, without two decades of gradualist reforms which ensured sufficient energy supply for China by using both TVE and SOE productivity, the government would not have been able to withdraw its investment and subsidies from uncompetitive SOEs and concentrate on building big business by concentrating its funds. Moreover, gradualist reforms ensured the existence of large SOEs, and it became much easier to reorganize them into large competitive modern corporations because they are state owned.

Gradualism is still required as more challenges are facing SOEs

China's state-owned sector is sharply reducing the level of employment, with a conservative estimation of 30 million workers being laid off between 1996 and 2001 (*Financial Times*, 21 November 2002). The urban non-state sector is increasing employment quite rapidly (from 21 million in 1995 to 35 million in 1999) (SSB 2000: 115). However, it is far from being able to absorb the large numbers made redundant from the SOEs together with the growing number of new entrants to the labour market. The urban unemployment rate (already eight per cent by 1998) is likely to accelerate in the period ahead, because the pressures upon SOEs by the big business revolution and China's entry to the WTO will be much fiercer than before.

As shown in Jixi's case, the standard of living of laid-off workers has greatly reduced, and strikes have occurred frequently. The nationwide situation is predicted to be even worse as the cost of shifting from the current pay-as-you-go to a fully-funded pension system by 2030 is 3,000 billion yuan, which will be a heavy burden for the government (*Financial Times*, 11 December 2002). At the same time, China also faces severe internal tension due to consistent rural poverty, increased social inequality, and widespread corruption. All of these indicate that a government-controlled approach, which is both cautious and well-planned, is still required in dealing with SOEs in the future. In other words, gradualism is still required in transforming the SOEs.

It must also be emphasized that transforming SOEs without dramatic privatization is not only no choice at all, because of its difficulties, but also an active choice because of the beneficial effects brought about by keeping the state ownership of some strategic important industries or enterprises. As shown in Chapters 1 and 5, the period since the late 1980s has witnessed for the first time the opening up of a truly global market place in goods, services, capital and highly-skilled labour. Even in the highly conservative coal sector, the world's largest three mining giants BHP Billiton, Rio Tinto, and Anglo American have all entered China and are actively seeking investment opportunities.

In this situation SOEs have understood the warning that they must both cut costs and restructure in order to benefit from economies of scale and to improve their competitive capacity. Restructuring, concentration, and consequent redundancies seem unavoidable for SOEs.

As shown in Chapter 5, the government has been actively building Shenhua Coal Corporation into one of the world's competitive corporations. By 2000, the total industrial value of the coal industry was 135 billion yuan, of which that of the SOE coal companies was 114 billion yuan, accounting for 84.2 per cent of the total (CEDR 2001:124). It is still a big question whether such valued SOE assets should and could be privatized in the near future. It could be impossible or at least much more difficult without taking the advantage of state ownership, which guarantees the highest speed and lowest transaction costs to build globally competitive indigenous companies so as to respond to the challenge from global rivals.

It should also be pointed out that it may be naïve to assume that the new creation of the State Asset Management Committee (SAMC) could solve SOEs' problems all at once or in a short time. The basic idea behind setting up the SAMC is to provide representation at the central, provincial and municipal levels to protect the state's shareholding interest in managing assets in SOEs, but not in managing SOEs directly. With the emergence of an officially legitimized ownership institutional framework, major obstacles to SOEs being taken over or declared bankrupt could be removed. Therefore, the establishment of the SAMC may indeed improve conditions for the further reform of SOEs.

However, numerous difficulties still exist. What could guarantee that the SAMC act as a real owner responsible for securing or adding to the asset value of the SOEs while achieving the appropriate degree of intervention in corporate governance? How are state assets to be valued? How are state assets to be allocated to the SAMC at the three different levels? And how is increasing local protectionism to be avoided once local governments manage their assets independently? Selling state assets is even assumed to be the best option for investors, as they acquire the potentially under-valued state-owned industrial capital (Chinaed 2003). All these difficulties or problems indicate that the SAMC does not yet provide the foundation for an easy and fast solution to the problems of China's SOEs.

It might be due to this reason that the 16th Chinese Communist Party Congress, while proposing to set up the SAMC, reiterated the importance of strengthening both SOEs and non-SOEs. This is exactly a continuation of the gradualist strategy China applied before: developing new ownership to compete with, and to force changes in, the enterprises that stay within state ownership. In terms of this perspective, gradualism is still current today and SOEs, especially the large ones, are still expected to play an important role in the immediate future.

Conclusions

A series of gradualist reforms has been carried out in China's coal industry since the 1980s, which partly contributed to the difficulties of SOEs: of which, the most fundamental gradualist reform – the dramatic development of TVEs – destroyed the monopolistic position of SOEs and consequently caused their financial difficulties. The force from rural underdevelopment created TVEs, whose attack was stronger than any 'shock therapy' to 'cure' chronic illness experienced by SOEs – lack of incentives to improve their competitive capacity and lack of understanding of profit making. Another gradualist reform – unchanging state ownership of SOEs – also contributed to their difficulties due to poor management, large social welfare spending, and so on. However, problems faced by SOEs were the result of many other factors. Gradualism therefore could be considered related to, but not be fully responsible for, SOE difficulties.

Gradualism was more favourable and feasible for China than the more precipitate alternatives followed (with distinctly mixed success) in other transitional economies. Transition poses great challenges for China, e.g. for a nation's financial capacity, for the traditional privilege and legitimacy, and for the available institutions, and laws and regulations. It would incur conflicting interests between gain and loss.[21] Moreover, development is the supreme challenge and transition must serve this ultimate purpose. Careless transition, under the influence of global economic and political change, might disadvantage a nation's development. Therefore, it appears more feasible and also more humane to deploy a cautious and gradual approach to reform.

This strategy is still current after two decades of gradualist reforms. As the result of changing world business practices, China should respond by taking a more strategic control of her most important industries by various means, including maintaining state ownership of them. Numerous problems and difficulties indicate that the SAMC does not yet provide the foundation for an easy and fast solution to the problems of China's SOEs. In conclusion, China deployed, and may still need to deploy, gradualism not only because the strategy is not as costly, but because it is suitable for China's situation of suffering from interactive challenges. In Stiglitz's (1999b: 36) word, '[w]e seek not perfection, [but] simply workable solutions, or, from the perspective of the transition, at least solutions that avoid disaster'.

Figure 4.7 Failing SOE mining bureaux during the transition

Top left: The headquarters of Jixi Mining Bureau, which, during the transition, was declining from the position of being one of the best SOE mining bureaux to becoming the largest loss-maker. *Top right:* Common working conditions in an SOE coalmine, which are among the worst in all industrial sectors. *Bottom left:* Facilities at the Xiaohenshan Mine, one of the four bankrupted coalmines of Jixi, now disused. *Bottom right:* Redundant coalworkers of the bankrupted Xiaohenshan Coalmine look for casual jobs in snowy weather, waiting for drivers of passing trucks or cars to ask them to load or unload.

5   Globalization

Building competitive coal corporations

> [A]fter entering the WTO ... we must face the fact that in capital and technology intensive industrial sectors, not only do we have a weak base and low starting point, but more importantly, our competitive advantage of labour intensive industry will cease. Although we already invest huge capital in a list of essential industries, to be honest, we are not sure whether they can grow up and can compete with global giants. If global giants control our market, most of our firms will be heavily disadvantaged, and our economic growth will likely be stopped by market constraints. Therefore, speeding the development of big business is vital for our survival, and also a strategy task which ensures the security of national industry .
> (SPC 1996: 6–7)

Introduction

In 1991 China officially designated 55 enterprise groups to trial the building of indigenous competitive groups, and then expanded this number to 120 by 1997. In addition, 300 enterprises were selected as key enterprises to be supported by the central government in 1996, and this number subsequently increased to 520 by 1999. These groups and enterprises are collectively known as the 'national team' and are considered as the mainstay in their respective sectors and the national economy as a whole (State Council 1991, 1997; CESRY 2001).

There is a list of active industrial policies to support these key groups and enterprises: extensive support from the banking sector; shelter from international competition behind a wall of protective tariff and non-tariff barriers; an independent accounting system which removed the barriers between different sectors, departments and regions; permission for the establishment of internal group finance companies; the granting of import and export rights; rights to establish international joint ventures, and rights to float a share of equity on national and international stock markets (CSSA 2002, 2001; Nolan 2001a, b; SPC 1996, 1999a).

After over a decade of effort, overall results are rather disappointing, despite important progress being made. By 2002 there were eleven Chinese companies in the *Fortune* Global 500, ranked by value of sales (*Fortune*, 22 July 2002). However, as *Fortune* comments, 'Chinese companies still account for less than two per cent of Global 500 revenues. ... All 11 companies remain state-owned, and overmanning is still rife. The oil giant Sinopec (No. 86) needs 937,000 employees to generate

$40 billion in revenues, while Exxon's 97,900 employees bring in $192 billion'. What *Fortune* does not point out is that China's most successful large firms, State Power, China Telecom, China Mobile, PetroChina and Sinopec, all operate behind high protectionist barriers and their operations are mainly confined to the domestic market.

China's industrial policy has failed to create globally competitive 'national champions' on a large scale. Should China continue to pursue an industrial policy when it seems so difficult for the industrial policy to build large globally competitive corporations? Should China focus instead on developing successful globally competitive firms within the global value chain of the world's leading 'systems integrators', headquartered in the high-income countries?

There seem to be more and more reasons for the Chinese government to give up its industrial policy. First, the mainstream view in development economics believes that undistorted prices, competitive markets and free entry for small firms, rather than big business and controlled markets, are the keys to development. Second, this period has seen truly global corporations being created, making the task of catching up for firms based in developing countries even more difficult. Third, even if China wished to continue its industrial policy, it would be much more difficult than before, due to joining the World Trade Organization (WTO).

Any given country's decision to implement an active industrial policy depends on both industrial needs and government capacity. This chapter, based on an in-depth case study of the Shenhua Group Corporation Limited (hereafter Shenhua), aims to provide evidence of the needs and possibilities of China's ability to build indigenous globally competitive companies. The second section shows that even in the outmoded coal industry, the global big business revolution has created a new breed of even more powerful super-large internationally competitive companies, which have created an enormous challenge for China. The third section shows that simultaneously by joining the WTO, China's weak SOEs and tens of thousands of TVEs will have to compete face-to-face with the global giants. The fourth section provides evidence that China is still capable of building a powerful big business provided a well-designed and active industrial policy is in position. A conclusion has been made in the final section.

The big business revolution and consolidation in the coal industry

The big business revolution

Instead of being a period of national champions and conglomerates, the business structure of advanced economies has been revolutionized since the late 1980s. Because of factors such as the collapse of communism, privatization, trade and capital flow liberalization, and new information technology, by the late 1990s there was a very high degree of firm concentration on a global scale in a wide range of sectors (Nolan 2001b: 97–122).

No. of deals   Value($billion)

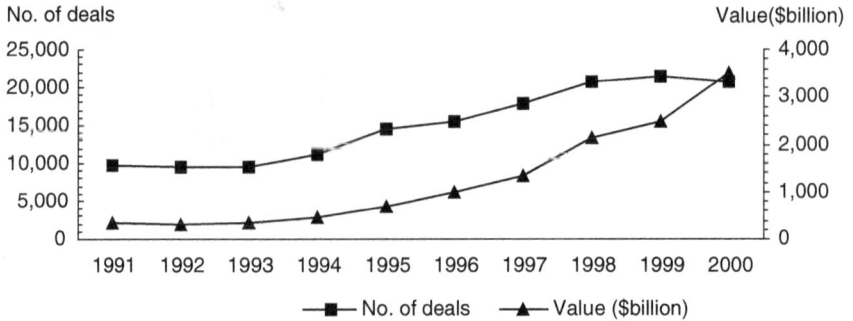

Figure 5.1 Ten-year global merger completion record 1991–2000

Source: *Mergers and Acquisitions* 2001.

This revolution has brought a dramatic growth in the business capability of major international companies, through concentrating on core business, enhancing brand names and massive spending on R&D and IT. Global mergers and acquisitions (M&A) rose from less than $300 billion in 1992 to around $3,500 billion in 2000 (see Figure 5.1). The concentration process was inexorable, permeating almost every sector. Firm after firm shed non-core business in order to focus on the areas in which the firm could compete globally, which has resulted in a small number of firms accounting for over a half of the global sales in numerous sectors, from the extremely high-tech aerospace sector, dominated by Boeing and Airbus, to the less high-tech coal sector whose international market is dominated by BHP Billiton, Rio Tinto and Anglo American.

The global value chain cascade effect on developing countries

The business revolution is relevant to this chapter because of its impact on developing countries such as China. Introducing a new concept of the global value chain cascade effect would be helpful in examining such an impact.

A value chain is a systematically coordinated production and delivery process, comprising all the stages leading ultimately to 'downstream' sale from the initial 'upstream' raw material processing. As business internationalizes, value chains are lengthening, extending across borders and drawing in more firms. However, global companies have also become more effective supply chain coordinators, with the aim of meeting customers' needs, minimizing costs, maximizing market share and profit, or strengthening strategic control.

Giant or core companies, acting as 'systems integrators', penetrate the global value chain deeply both upstream and downstream. They are closely involved in business activities that range from long-term planning to meticulous control of day-to-day production and delivery schedules. This has created an explosive 'cascade' effect on suppliers, causing rapid restructuring among first-, second- and even third-tier suppliers, working across sectors and between developed and developing countries. For example, the intense competition and consolidation among automobile makers forced the consolidation of their supply companies, including those providing components and steel; and the liberalization and high-speed consolidation in the global steel and electricity markets promoted consolidation among their suppliers, including those providing iron ore and coal. In most sectors of the entire value chain, there is an ever-larger sphere of co-ordination and planning of the entire value chain orchestrated by the core systems integrators.

As a result, the boundaries of the large corporation have become blurred. The more indigenous firms join the global value chain as either suppliers or customers and the more multinational companies (MNCs) relocate in developing countries, the deeper and broader the global value chain effect will impact developing countries through the unitary planning of the 'system integrators', i.e. powerful MNCs. In other words, the national interests of the host countries may become downgraded in relation to corporate interests, and policy makers in developing countries may well have to consider carefully both the role of indigenous firms within the global value chain and the national economic security.

While the positive effect that globalization and large FDI flows brought to developing countries is highlighted, a crucial fact has been neglected: the regional distribution of firms that lead the global big business revolution has been hugely uneven. The massive disparity in the regional base of the world's leading firms is a deeply sensitive issue in international relations. Failure to recognize this disparity and the way, in many respects, that it has become more unequal makes international economic agreements between developing and advanced economies more difficult to negotiate and implement. Table 5.1 shows that regions containing a small fraction of the world's population have massively dominated the global big business revolution. As the world moves into the new millennium, developing countries are enormously disadvantaged in the race to compete in international big business. The starting points in the race to dominate global markets are highly uneven.

Consolidation in the coal sector during the big business revolution

People typically associate the concept of big business with sectors such as the aviation and automobile industries, and regard the coal industry as a declining sector. However, the big business revolution is affecting even this sector. Compared with the institutional changes under way in most other industries, the coal industry still lags behind in terms of global consolidation. However, the speed of consolidation is still remarkable. In the late 1990s, M&A in the coal sector grew to record levels, reaching $17 billion in 1995, rising to a new peak of $26 billion in 1998. In 2001, the process scaled new heights. The BHP/Billion merger alone was valued at $31.1 billion (*FT,* 24 March 2001; *The Sunday Times*, 25 March 2001). Following the rapid growth of M&A, the future of the international coal trading industry is seen as an oligopoly with interests in the three main exporting centres, South Africa, Central America and Australia. The market has been controlled by a small number of powerful international mining companies: Rio Tinto, Anglo American, BHP/Billiton with over 62 per cent of the total

Table 5.1 Dominance of firms based in high-income countries of the big business revolution

	Population		GDP, 1997[1]		GDP, 1997[2]		Fortune 500 companies, 1998[3]		FT 500 companies, 1998[4]		Top 300 companies by R&D spend 1997		Stock market capitalization 1997	
	Billion	%	$billion	%	$billion	%	No.	%	No.	%	No.	%	$billion	%
High-income economies	926	16	23,802	80	21,091	57	474	95	484	97	298	99	18,452	91
Low- and middle-income economies	4,903	84	6,123	20	15,861	43	26[5]	5	16[6]	3	2	1	1,725	9

Sources: World Bank 1998: 190–1 and 220–1; *Fortune*, 2 August 1999; DTI 1998: 70–80; quoted from Nolan 2001: 17.

Notes
1 At prevailing rate of exchange.
2 At purchasing power parity dollars.
3 Ranked by sales revenue.
4 Ranked by market capitalization.
5 Of which: Korea = 9, China =6, Brazil =4, Taiwan =2, Venezuela =1, Russia =1, India =1, Mexico =1, Malaysia =1.
6 Of which: Hong Kong =7, Brazil =2, Taiwan =2, Singapore =1, Mexico =1, Indio =1, Korea =1, Argentina =1.

internationally traded coal in 2001 (Table 5.2). Each of them is still actively pursuing a policy of M&A.

Competitive advantage in the global coal industry lies primarily in size and management skills. Size enables international outreach and the construction of a set of low-cost coalmines around the world. This reduces risk and ensures location within, or close to, each major market. Successful global firms also have the financial resources to purchase mines, invest in exploration and evaluation of resources, negotiate with governments, deal with complex land rights and environmental issues, apply best practice across the whole mining business, integrate their global mining operations within a single global marketing system, and thus they are able to force down costs more successfully than smaller local companies. Most global coal companies are not simply coal companies. Rather, they manage diversified operations focused on mining products. Their size and capital advantage enable them to adjust themselves to business shift. Size and consolidation also ensures that coal companies, or the mining sector as a whole, can increase its bargaining power with other sectors.

The BHP/Billiton merger was pursued precisely in order to enhance the firm's competitive advantage:

> The companies balance each other well, with an exceptional breadth of assets and capabilities which have taken many years to develop! This merger brings together some of the world's finest mining, metals and energy assets under a dynamic and unified executive team. Few, if any, of our competitors will be better placed to serve the commodity requirements of our diverse customer base. The financial strength, international scope, and enhanced project skills of the combined group should bring major new growth opportunities internationally.
>
> (*FT*, 18 March 2001)

Two years later, with these advantages, BHP Billiton has become an industry leader or a near-leader in aluminium, metallurgical coal, seaborne steaming coal, copper, ferro-alloys, iron ore and titanium minerals (BHP Billiton 2003).

In addition to size and management skill, brand is also crucial. Here brand has a broad meaning, including service quality, supply reliability, trust, unwritten norms and rules, and even cultural familiarity which are embedded in the global value chain. For example, although the price and quality of Shenhua coal of China are competitive with Australian coal, Shenhua's director admitted that Australian coal is more competitive in the global market because it can meet customers' demands through its global service network and high skills in blending coal[1] (*pei mei*) to supply customers' precise requirements. Shenhua obviously lacks experience in these crucial aspects of coal supply.

In 2002 and 2003 the world economy was in recession. However, the coal business, exceptionally, was enjoying prosperity. Following its acquisition of the Glencore International AG's Australian and South African coal business in the first quarter of 2002, Xstrata was fast becoming one of the world's largest export

Table 5.2 Major indices of the world's and China's leading coal companies in 2002

	Coal output (Mt)	Global traded coal[1]	Market capitalization[2] ($ billions)	Employees[3]	Turnover ($ millions)	Net earnings ($ millions)	Main products
Rio Tinto	149	149 (26.7%)	13.2	36,141 (5,550)	10,828	651	Industrial minerals, aluminium, iron ore, copper, gold, coal
BHP Billiton	120	120 (23.7%)	32.9	51,037 (18,496)	16,168	1,900	Aluminium, ferro-alloys, nickel, coal, base metals, oil and gas, iron ore, coal, steel
Anglo American	65	62 (11.9%)	21.8	204,000 (10,000)	15,449	1,563	Platinum, diamonds, forest products, gold, industrial minerals, coal, base metals
Shenhua	77	20	11.0	90,000 (9,716)	3,110	382	Coal, electricity, (oil products by 2005)
Yanzhou	38	14	2.8	82,000	1,361	147	Coal, (limited production of coal water mixture, urea, and methanol products)

Sources: Annual Reports, various years for each company; Osiris 2003.

Notes

1 Data for the top three global companies are for 2001. Global coal trade was 625 Mt in 2001. The figures in brackets refer to the percentage shares of the global traded coal for each company.

2 For China's coal companies, the figures refer to their assets. The capitalization for the three global companies, the figures refer to their assets. The capitalization for the three global giants refers to data at 31 December 2002, 30 June 2002 and 31 December 2002 respectively.

3 The figures in brackets refer to the employees dedicated to coal business.

thermal coal producers, with a capacity of 70 Mt of coal output per year (Xstrata 2003).

Strengthening China's industrial policy to meet the challenge

Joining the WTO will force China to 'dance with wolves'

The last section analysed the global big business revolution, which has created giant international firms. During the same period China was making great efforts to be admitted to the WTO. In November 1999, China and the US signed an historic agreement under which the former would be permitted to join the WTO. Through this decision China was bowing before the extraordinary force of oligopoly capitalism. Under the terms of the agreement, China was to dismantle almost the entire range of mechanisms of its industrial policy which supported the growth of large indigenous corporations in the past. The US–China agreement is the most detailed signed by any country on its entry to the WTO to date. The agreement in itself constitutes a massive programme of economic system reform. Nine hundred Chinese laws will need to be changed and/or adapted for China to enter the WTO.

The Chinese government was well aware of the deficiencies of its large firms in competition worldwide, and this is clearly reflected in the warning from the State Planning Commission (SPC) as early as 1996, quoted at the beginning of this chapter. As the SPC pointed out, after joining the WTO, China's domestic firms which must face the challenge of global competition consist of thousands of large but weak state-owned companies and tens of thousands of small and weak TVE companies. During the reforms of the last two decades creating an internally integrated market and encouraging the emergence of real firms was regarded as a more urgent task than meeting the challenges of international competition.

Before reform began in 1978, four decades of implementation of the central planning system had led to China having no firms in the conventional sense. Instead there were hundreds of thousands of '*da er quan, xiao er quan*' (large and complete, and small and complete) plants, which lacked specialization, were dispersed, small scale and limited by sectors and regions. Thus, cutting away excessive vertical ties and removing sectoral and regional protection, while at the same time building horizontal linkages between producers of similar products was necessary for a greater specialization in production which, in turn, could foster economies of scale. In the absence of capital and labour markets, the development of supra-departmental enterprise groups was first expected to break through these regional and departmental barriers (SPC 1996: 5–6). China's economic miracle during the past two decades has depended mainly on the large investments of capital, labour and natural resources, and the rapid development of tertiary industry. It has been realized by more and more policy makers and reformers that China's new growth strategy should focus on achieving better economies of scale and better technical progress rather than only investing more resources and labour (from *cufang* to *jiyue*) (SPC 1996: 6). Building large modern corporations is expected to create new sources for growth for the entire economy.

China's coal industry offers a prime illustration of how severe are the problems which resulted from the lack of economies of scale domestically and from the global big business consolidation. It is well known that coal has held a special position in China, in terms of its dominant share in the primary energy production and consumption, in generating electricity, and in producing steel and other chemical products. Such an important industry was nonetheless, as shown in the previous chapters, the principal loss-maker among China's industrial sectors until 2001 (except in 1997). A mass of evidence demonstrates the low levels of efficiency and financial difficulties of the sector; problems which are rooted in severe diseconomies of scale and scope.

Table 5.3 reflects the severe degree of diseconomies of scale in China's coal industry: before the government's closure policy started in 1998, the average output per company was only about 17,101 tons. After the closure policy, the average output was increased to 50,000 tons by 2002. The market share of the largest 10 coal producers increased from 13 per cent in 1997 to 21 per cent in 2001; this compares with the US where the top five coal producers accounted for 51 per cent of total coal production in 2001 (NMA 2003). Among China's coal companies, 'none has significant market share and none could influence market competition' (SCA 2000: 5). Moreover, few coal companies have dedicated rail, port and power plants, leading to high transaction costs and placing coal companies in a very disadvantageous position.

This sort of diseconomy of scale and scope has inevitably caused excessive competition, extremely scattered capital and technical investment, a chaotic coal market, depressed coal prices, low productivity, and large-scale losses. More importantly, dispersed capital and excessive competition also obstruct techno-logical progress and the capacity among enterprises to merge. Excess competition

Table 5.3 Changes in size and structure of China's coal producers 1998–2002

	Number of producers (Mt)		Total output		Average output per producer (tons)		% of national output	
	1998	2002[1]	1998	2002	1998	2002	1998	2002
National total	72,042	27,666	1,232	1,393	17,101	50,000	100	100
Large/medium SOEs (>450,000t)	448	283	449	712	1,000,000	2,500,000	36	51
Small SOEs (30,000–450,000t)	1,794	2,383	172	263	96,000	110,365	14	19
TVEs/small SOEs (<30,000t)	69,800	25,000	612	418	8,800	16,720	50	30

Source: SCA 2003; SSB 2001: 48, 80; Yan *et al.* 2000: 28.

Note
1 Data for large/medium SOEs and small SOEs were for 2000.

prevents most SOEs from producing at full capacity thus making losses inevitable, and mergers impossible (for lack of strength). Moreover, too many producers from various regions, provinces and government departments, local and sector interests block mergers and acquisitions among themselves.

While China's coal industry suffers from diseconomies of scale and scope domestically, the international coal giants have recently gained strength through an unparalleled process of consolidation (as seen in the last section), and are accelerating their entry into China causing the threat to indigenous firms to be more intense than expected.

During the relatively short period of 2002–3, all the global mining giants including BHP Billiton, Rio Tinto, Anglo American, and CVRD have not only entered China, but have completed preparation work and started to 'take action'. This action is based on the belief that China is the 'final frontier' for their business success and that China's coal market is too large to neglect. One confidential report from a global mining company stated: 'Foreign players need to secure their positions in the market early to ensure access to attractive opportunities, as there will only be a selective few players that will emerge as industry leaders'. Concurrently, the Chinese government, in November 2002, granted foreign mining companies the right to extract coal in China. The world's leading companies are actively engaged in establishing joint ventures in order to save time and costs in mine and rail building, and to acquire local market knowledge. The government has established few special conditions for foreign firms in respect of tax and resource access. Therefore, joint ventures in this sector will truly compete with domestic companies on a level playing-field in both the home and international markets.

The threat from the MNCs is reflected not only in the fact that the international mining giants may compete in China's domestic and export markets, but also in many other aspects. Although the reduction of tariffs on coal (from 6 per cent in 2001 to only 3 per cent by 2005) as a result of China's joining the WTO was assumed not to be significant, however, the liberalization of non-tariff regulations was believed to result in more imports of coal and coal equipment. In 2002 coal imports into China increased to 10 Mt (from 2.49 Mt in 2001) which led to the reduction of the domestic Qinghuangdao benchmark price, to the detriment of the domestic coal producers. In addition, both tariff cuts on imported oil and gas, as well as environmental concerns, may result in more oil and gas being imported so as to attract former coal users. Finally, as a result of the cascade effect of the global value chain, as explained above, the relocation of the MNCs in manufacturing may well impact on domestic coal companies on quality, price, and service.

To avoid the fate of being merged, acquired or closed, China's coal companies need to run their own large efficient mining businesses in order to achieve economies of scale, to produce a competitive price and high quality of coal, and to withstand intensified global competition.

Building big business by industrial policy rather than by market forces

If China is to develop its own globally competitive big business, should it wait for firms to emerge spontaneously from free market competition, or actively build them through well-designed industrial policy? According to the mainstream view, the market mechanism should allow free entry and exit for small firms, with the 'winners' emerging naturally through market competition. In the absence of such a free competitive market, as in the case of socialist economy, privatization is regarded as the primary task of the 'transition'. Restructuring should be left to the new private owners. State-led industrial restructuring, including building big business, is considered to be unnecessary and undesirable.

However, the case of China's coal industry has demonstrated that it can be impractical and costly to leave restructuring to the new owners after privatization in transitional economies. In the past two decades China's reforms have made great progress in establishing a competitive market. These reforms have created tens of thousands of new competitors for large SOEs. The free entry of TVE coalmines destroyed the SOEs' monopoly rent. The coal price was liberalized in 1994. Government subsidies were reduced and a warning has been given that they will cease completely. Bank loans to SOEs in financial difficulties have become difficult if not impossible to obtain. However, after two decades of liberalization, no big Chinese coal groups had emerged as a result of market competition. In contrast, there is strong criticism of the government's weak industrial policy, including its failure to control new entrants to the sector, failure to build large-scale coal-electricity corporations, and failure to support export-oriented coal companies with market guidance and lower tax and fee requirements. These failures have been identified as the major cause of the sector's excessive competition, coal surplus and financial difficulties.

The case of the Jixi Mining Bureau examined in Chapter 4 also demonstrates that rapid privatization in China's coal industry was not feasible in the specific institutional setting of its transitional economy. In China's coal industry, although no large SOEs have been privatized, 50 large SOE coal companies or mines were forced into bankruptcy in 1999 and 2000, and a further 220 SOE coalmines will be going bankrupt by 2005 (CCC 2000). The critical problems encountered include the absence of laws and regulations and insufficient financial capacity to pay the huge bankruptcy costs, The number of redundant workers grew rapidly, but there was no social security system to deal with issues such as pensions and unemployment benefits.

Most significantly, many industries are too important for the government to allow them to develop without industrial policy guidance. In the case of the coal industry, the strategic significance is reflected in the fact of China's severe shortage of oil and great reliance on coal. Coal will become even more crucially important after the government implements the coal liquefaction project (see below). Keeping state ownership of the most important coal companies might be a major advantage for the government in controlling and guiding the industry to serve the nation's strategic energy needs.

The Shenhua Group: can China build competitive indigenous big business?

The last two sections have shown that if China is to be successful in building her own globally competitive big businesses, it will require extensive support from a state industrial policy. This section uses the example of the Shenhua Group to shed light on China's capacity to build large globally competitive firms.

China's ambition to build big businesses in the coal sector has been pursued explicitly since the 1990s. Although this sector is still highly fragmented, it has also made significant progress as a direct consequence of the government's active industrial policy. As one of the key trial enterprise groups, Shenhua has been built from nothing in 1985 to its current position as one of the most important competitive coal corporations in the world. The speed of development can be seen from Table 5.4. The degree to which the government is able to utilize China's competitive advantages successfully to support such a dramatic rise in Shenhua will provide a significant indicator of its capacity to implement its industrial policy of building large firms, known in China as the 'national champions'.

China's advantages in building big business in the coal industry

First, China is an immense country with abundant resources. Although the per capita levels of these resources in proportion to its 1.2 billion population is small, in many sectors the national total is huge. The government can make use of the country's abundant and diverse resources to support the growth of its big business. In Shenhua's case, the government gave them a mandate to extract coal from the Shenfu Dongshen coalfield (on the borders of Shaanxi and Inner Mongolia), whose total recoverable reserve is as much as 223 billion tons, with estimated reserves of 1,000 billion tons. This compares with a national total of 1,100 billion tons of recoverable coal reserves. Shenhua is therefore planned as China's most important future coal base as north and central China's coal is substantially depleted.

Second, China's state-owned businesses still occupy a central position in the economy, and this is usually seen as a disadvantage by privatization advisors, giving the government a powerful weapon for restructuring industrial assets. So as to follow the model of the world's leading coal producers closely, Shenhua was designed with its own railway line to ship coal from the Shenfu Dongshen coalfield to a dedicated port facility. It was also assigned dozens of power plants from the SPC to consume millions of tons of coal from its own coalfields.

The construction of Shenhua based on this design required approval and coordination from more than four ministries and six local governments, responsible either for the coalfield, local land, or infrastructure.[2] As the State is the ultimate owner of all of these properties, and also the ultimate administrator of all levels of

Table 5.4 Output, sales revenue and profit for the Shenhua Group from 1998 to 2002

Year	Coal output (Mt)		Coal sales (Mt)		Coal export (Mt)[2]	Rail freight (Mt/km)	Shipped through port (Mt)[3]	Electricity generation (M kwh)[4]	Sales revenue (million yuan)	Profit (million yuan)
	Without FWB:[1]	With FWB	Without FWB	With FWB						
1998	7.13	13.75	9.24	14.69		2,395		330	2,422	52
1999	11.20	18.40	14.12	20.42		3,669		340	3,879	44
2000	24.12	37.54	25.22	38.59	7.5	10,554		16,190	14,099	176
2001	35.86	52.74	40.20	56.02	18.0	19,167		20,490	19,900	1,573
2002	45.02	77.33	57.10	77.73	20.0	31,966	16.53	23,860	25,720	3,158

Sources: Shenhua Group Annual Reports, from 1998 to 2001; Shenhua Group, memos, various years; Chen, 2003.

Notes
1 FWBs = Five Western Bureaux.
2 The data in the last six columns are for the Shenhua Group as a whole.
3 The port started to operate in 2002.
4 Since 1998 Shenhua has acquired several power plants, established as 'Guohua Electricity Corporation Ltd, Shenhua' in 1999, to use its own coal production. This enabled Shenhua's total power generation to rise to 122,800 million kwh in 2000.

government departments and local governments, it was therefore able more easily to grant Shenhua the use of all the properties and to ensure better coordination among all government departments. It would have been much more difficult for a private individual to coordinate these complex provincial, regional and departmental interests. So far this is the only coal group to achieve this sort of economy of scope with correspondingly low transaction costs.[3]

Third, big business needs large investment. Lack of capital is a major constraint for developing countries trying to build large, capital-intensive enterprises. Although China is relatively poor compared to other well-developed countries, it can concentrate dispersed funds for important collective ventures. The high level of savings[4] and foreign direct investment (FDI) can help meet the huge investment requirements. In Shenhua's case, the central government granted the Shenhua project preferential loans of over $9.2 billion during its construction period from 1985 to 2005. This made it the third largest investment project in China and one of the world's largest overall. This compares with a figure of $1.3 billion for China's total investment in the coal sector in 2002 (CSD 2003: 57). The government also made use of local funds and invited Hebei province and the Ministry of Communication to invest a small part by granting them a minor shareholding each. This will also help to reduce any tendency to local or sectoral protectionism against Shenhua.

Fourth, possessing advanced technology is another determinant of the success of big business. China's coal industry is relatively backward by comparison with the rest of the world, but this did not prevent the government from importing advanced equipment for Shenhua. In addition to having more than 98 per cent of its equipment imported, Shenhua has installed a reliable automated despatching system for underground mines, thus making substantial labour savings. This illustrates very significant support from the government, which not only provided huge investment, but also, exceptionally, allowed Shenhua to replace labour with imported equipment. This provoked a big debate on whether China should continue to make use of its comparative advantage of cheap labour and avoid possible social disruption caused by the large numbers of redundancies resulting from the use of advanced equipment, or whether it should instead seek to improve productivity by using advanced technology and equipment as early as possible in order to compete with the giant coal companies. In the end, the latter view prevailed, demonstrated by the fact that Shenhua's principal coalmine – Daliuta – has established itself as one of the most advanced in the world, with the world's highest productivity record – 118 tons per employee per day – in 2002 (see Table 5.5). Daliuta's productivity was recognized as the world's highest at the World Long-Wall Conference held in the USA in June 2001 (Shenhua Annual Report 2001).

It should be noted that other mines in Shenhua have also followed this 'high output and high efficiency' philosophy and been built with very advanced technology, equipment and management. In 2002 the newly built Yujialiang Mine produced 10.59 Mt with 248 employees, a productivity of 122.7 tons per employee per day.

Table 5.5 Some indices at the Daliuta Mine of the Shenhua Group, 1996–2003

	Total cost (yuan/ ton)	Tons/ employee/ day	Tons/ employee/ day (for long-wall)	Output/ year (Mt)	Employees	Average annual wage (yuan)
1996	54.43	12.31	60.37	2.31	706	–
1997	51.89	15.08	63.68	2.86	688	–
1998	51.06	19.64	127.43	3.27	439	–
1999	47.86	41.54	250.02	5.09	412	–
2000	44.96	83.00	516.00	8.00	350	25,000[3]
2002[1]	46.68	117.81	595.31	15.27	495	34,500
2003[2]	47.84	124.67	661.14	19.00	495	>34,500

Sources: Daliuta Mine 2000; Shendong Coal Ltd 1999; 2003.

Notes
1   The Daliuta Mine was merged with the Huojitu Mine by 2002.
2   Data for 2003 as recorded up to June 2003, and estimated for the rest of the year.
3   By 2000 the average annual wage per employee at Shendong Coal Ltd (including Daliuta) was 25,000 yuan, which increased to 34,500 yuan by 2002. This compared with that of about 5,000 yuan at Jixi Mining Bureau in 2000.

Fifth, a core business with high potential demand is another key to the success of big business. Shenhua's high-quality low-sulphur and low-ash coal is in great demand. In addition, the government also supports Shenhua as a pioneer in the development of coal liquefaction. Coal liquefaction is a technology through which coal is transformed into oil.[5] This is currently not significant, but potentially extremely important for both China and the world's economy. In 1993 China became a net oil importer and it is forecast to import 100 Mt of oil in 2010. In line with the strategy of not allowing foreigners to control its oil industry, the government strongly supports Shenhua as the operator of the programme by awarding almost all the nation's funding for 'substituting coal for oil' to Shenhua.[6] As the cost may be controlled to be below $20 per barrel, the market for Shenhua's oil products will have a promising future. Chairman Ye was confident that by 2020 Shenhua should be more an oil-producing company than a purely coal-producing company.

Finally, under China's centralized and hierarchical political system, millions of officials and public servants are still appointed by the organizational departments of the Chinese Communist Party (CCP). The central government appointed Ye Qing to be Shenhua's Chairman, Chief Executive and Party Secretary. For over a decade, from 1986 to 1998, Ye was deputy director of the State Planning Commission (SPC) in charge of energy and communications. This background determined his suitability to manage Shenhua because the structure of this conglomerate required both the authority and ability to coordinate all sectors in so many regions. In fact, Ye's powerful leadership in Shenhua after 1998 was the key to its success.

Shenhua's performance was the result not only of the government's support, but also of its own management capabilities. When Ye arrived in 1998, mine construction was still being concentrated on. As the coal market was already in surplus, and because of the high transport cost of Shenhua's coal, Ye was determined to cease the mine construction and accelerate construction of its dedicated rail and port facilities. The effect of this shift in strategy was very clear: Shenhua's railway and port construction was completed in 2001, two years ahead of schedule, saving 40 yuan per ton of coal in freight costs. This greatly improved Shenhua's competitive capacity.[7] Moreover, a key reason for its low cost of production was that Ye Qing was able to achieve the merger of the two formerly parallel companies, the Shenfu and the Dongshen Coal Companies, which ensured that Shenhua was able to cut unnecessary costs, make better use of its facilities and staff and respond more quickly to market changes. Ye Qing also implemented a succession of internal management reforms including a wage contract system, cost control contract system, quality control system, and asset and equipment utility responsibility system, moving Shenhua closer to a modern enterprise system.

China's benefits from building big business in the coal industry

We have seen that China's objectives in building big business are to restructure industry in order to achieve economies of scale and scope domestically, and to improve its capability to compete with the global giants. Shenhua has made obvious progress in respect to both of these objectives.

The Chinese government has been trying since 1998 to close TVE coalmines in order to pursue a policy of enforced economies of scale, a policy which has met strong resistance. However, Shenhua was able successfully to 'close' some TVE mines: Shenhua leases the right to extract coal from the TVE mine at a fee of 10 yuan per ton. Local governments keenly supported the deal as they were able to receive much more income than otherwise. The potential benefits from such a deal not only enabled local governments wholeheartedly to close TVE coalmines, but also to fundamentally reduce rural poverty. Shenhua's leasing of two TVE coalmines in Baode county in Shanxi province gave Baode 32 million yuan in lease fees,[8] which was considerably greater than Baode's total government revenue of just 24 million yuan in 1999. With Shenhua's production expansion in Baode, the latter's revenue has increased from 27 million yuan in 2000 to 68 million yuan in 2002 and an estimated 100 million yuan in 2003. These increases has assisted Baode in eliminating longstanding poverty (Ye 2003).

Shenhua's railway not only contributes high direct income to the local area, but it also plays an important role in providing high-quality infrastructure for the export of local resources outside the western region. Shenhua's newly built railway lines between Baotou and Shenmu, Shenmu and Shuozhou, and Shuozhou and Huanghua enable neighbouring producers for the first time to transport coal, cement and peanuts from their remote mountainous regions, greatly stimulating the local economy and assisting in poverty reduction.

Shenhua also made great progress in pursuing its economies of scale and scope (see Table 5.4) as shown by its core coal business and by its supporting rail, port, power and future coal liquefaction businesses. While its dedicated rail and port services saved 40 yuan per ton of freight costs, the power plants which it received (from the SPC) and those which it later acquired, which may have a generating capacity of 9 million kwh, will use 18 Mt of Shenhua's coal. The planned coal liquefaction plant is expected to consume 100 Mt of coal by 2020.

Internationally, oil prices rose dramatically in 2000, which led some global oil customers to turn to China's coal. Oil price rises also led to higher freight rates which discouraged Australian and South African coal producers from exporting to the Asian coal market. Seizing this opportunity, China's coal exports increased from 38.58 Mt in 1999 to 58.84 Mt in 2000, and to 85 Mt in 2002 (Figure 5.2). Of China's four coal exporters, Shenhua is the growing exporter (Figure 5.3). It is its low price of coal and excellent marketing capabilities that have enabled Shenhua to have an increasing share of the coal export market with the most powerful global suppliers such as BHP Billiton. Shenhua's coal exports soared from 2.12

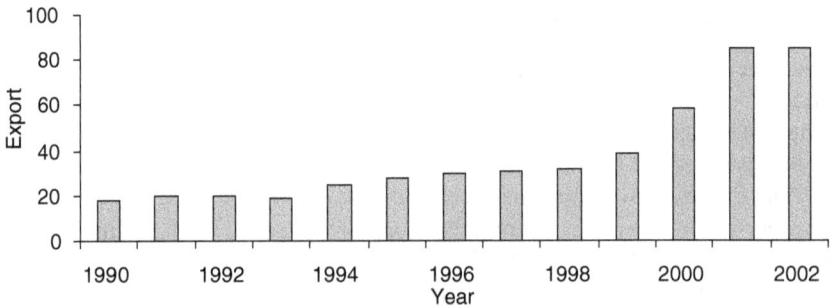

Figure 5.2 China's coal export 1990–2002

Source: CNCIEC 2002.

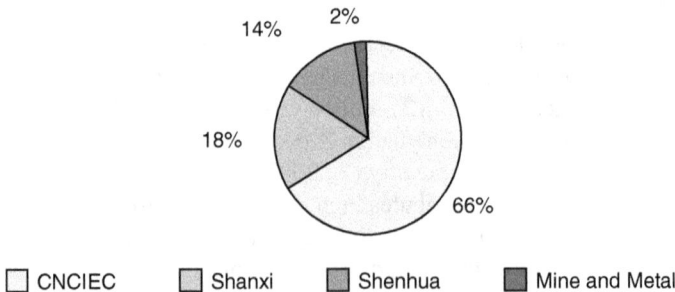

Figure 5.3 Shenhua's market share of China's exported coal in 2001

Source: CNCIEC 2002.

Mt in 1999 to 7.35 Mt in 2000, up 247 per cent; then to 16 Mt in 2001, up 118 per cent, and to 20 Mt in 2002.

Compared with the global coal giants, Shenhua is currently still small in terms of several factors such as capital and profit. However, its competitive potential is large. It has promising ambitions. The most updated plan for Shenhua is that by 2020 coal output should reach 400 Mt, of which 100 Mt will be for coal liquefaction and another 100 Mt for its power plants; oil production should reach 30 Mt through coal liquefaction projects, and electricity generation should reach 60 Mkwh. Together with its railway and port which are already in operation, Shenhua has attracted numerous companies who are keen to cooperate or merge with it.

Apart from the advantages analysed above, several features of the global coal industry could assist Shenhua in catching up with or even passing the global giants. These include the fact that the level of consolidation in the industry is still low (no single company accounts for more than around five per cent of total global coal production); the industry is relatively competitive, compared to sectors such as aerospace, oil and petrochemicals; leading global companies in the mining sector do not rely mainly on research and development for their competitive advantage;[9] firms in this sector do not establish competitive advantage through large outlays on building a global brand name; the capital goods necessary to operate large modern coalmines are easily available from specialist mining equipment companies, without any restriction on technical transfer; and finally, labour-intensive operation and geographical location provide China with advantages over Australia and Indonesia in exporting to the world's biggest coal-importing markets – Japan and Korea.

Constraints on building big business in the coal industry

Shenhua's immense potential cannot disguise the huge difficulties that China's government still faces in building its big business. Shenhua itself illustrates many of these problems and constraints.

First, the fundamental problem is rooted in China's low level of development, which determines that Shenhua cannot purely focus on competing with global giants or making a profit, but has to share social responsibility with the government, to look after loss-making SOE mines, and to compete with tens of thousands of TVE mines struggling for their survival.

In 1998 the key SOE coal bureaux were decentralized to the provincial level with the abolition of the Ministry of Coal Industry. However, since many SOE coal bureaux were heavy loss-makers, local governments either refuse to accept responsibility for them, or negotiate with central government for subsidies. The central government's decision to force Shenhua to take over the Five Western Bureaux (FWB) of Inner Mongolia in 1998 was precisely because the central government could not provide the subsidies requested by Inner Mongolia government as a condition of accepting the five, mainly as three of the bureaux were making heavy losses.[10] As a result, Shenhua's entire configuration was transformed overnight and its competitive capacity drastically limited. The FWB brought with them about 60,000 employees, huge liabilities, and outdated facilities and technology, compared with Shenhua's 22,000 employees,[11] substantial profits, and world-class equipment and technology.

The central government failed to control the establishment of TVE coalmines because it hoped they would contribute to poverty reduction. Although TVE mines did contribute greatly to local development, their explosive growth became a major source of China's coal surplus. The surplus market is, in turn, a major constraint on the market expansion of the big coal groups.[12] In addition, local governments are naturally inclined to protect smaller local enterprises, while the large groups are usually answerable to the central government in terms of loans, land use, infrastructure facilities and other services.

Second, the capital resources of China's big groups are too small. Although Shenhua's total investment was planned at $9.2 billion, all of which was loans, it had a capital fund of only 2.5 billion yuan ($302 million) in 2000. Shenhua has to repay loan and interest of 31.4 billion yuan from 2001 to 2010, at average 3.14 billion yuan each year, and 4 billion yuan in the peak year. Shenhua's average coal production cost was 67.89 yuan per ton in 2002, of which one-third was the cost of repaying loans and interest. This severely damages the competitive position of Shenhua coal and handicaps its further expansion. In addition, because of the severe shortage of funds, R&D spending in Shenhua is so small that it doesn't even provide a figure for its annual expenditure.

Third, Shenhua is constrained also by other weaknesses that are common to all SOEs such as the social welfare obligations, huge debts, overstaffing and frequent intervention from the government.[13] Moreover, Shenhua's achievements appear to owe a great deal to the qualities of its leader Ye Qing. However, in July 2003 Ye Qing retired and Chen Biting's appointment was announced.[14] It was announced that great changes were taking place in Shenhua following the leadership changes, including the suspension of the coal liquefaction project. It is far from certain that Ye will be replaced by an equally effective leader.

Fourth, and probably most importantly, the government lacks a comprehensive industrial policy to support the building of large groups such as Shenhua. This is reflected first in the lack of a unitary authority to design and control an industrial policy to guide the development of large indigenous firms. The administrative system of China's coal industry since the decentralization in 1998 is still ambiguous and ineffective. The policy problem is also reflected in the lack of coherence between detailed policies in other government departments and industrial sectors. For example, most of Shenhua's finance comes from the central government, which is allocated through China's State Development Bank (SDB). As the state's policy bank, the SDB should allocate loans to Shenhua in a timely manner and according to the budget plans once the State Development and Planning Commission has approved Shenhua's application. However, in recent years, China's financial system has been moving toward a more commercial basis, so the SDB increasingly adopts strict policies to supervise Shenhua's use of funds, and has frequently suspended or delayed loan allocations. This is, in principle, in order to prevent loss of state assets, but does in fact cause losses because of project delays.

The policy problem, as mentioned, also touches on another issue: how to control foreign investment in the coal industry. This will be very important for Shenhua in determining its development strategy, both in terms of how it will build its strength and on its market expansion plans. The government does not have a mature view on how to treat international coal giants and foreign investment in this industry on vital questions such as whether China should allow their entry, the procedures for establishing joint ventures in the coal industry, and taxation and royalty payments for foreign-owned coalmines.

Shenhua's problems are not unique, but reflect typical problems for all the large groups in China, as the SPC has acknowledged (1996, 1999a, 1999b). Obviously, in order to know whether China can build its own indigenous big business on a large scale, one needs to examine whether the government can eliminate or reduce these constraints. The answer remains uncertain. The government cannot neglect old SOEs and TVEs. The SOEs have large workforces for whom they still provide social welfare provision, so these enterprises cannot be eliminated. TVEs are supposed to have fewer problems than SOEs, but their condition is even worse from another angle, because SOEs are fiercely protected by the local governments.

Building big business beyond Shenhua's experience

Shenhua is just one of the successful cases in the process of China building its own big business. Before reaching a conclusion, it is necessary to have an overview of the big business in China as a whole. In general, large enterprises performed much better than medium-sized and small enterprises in terms of turnover and profits (e.g. see Jefferson *et al.* 2003). For example, in 2000, 515 key state enterprises made profits of 225.3 billion yuan, accounting for 98 per cent of total SOE profits (i.e. the other SOEs in total made almost no profits) (SETC 2001). In 2002, 510 key state enterprises made profits of 280 billion yuan (SETC 2003a), accounting for 104 per cent of total SOE profits in 2002 (i.e. the other SOEs were making losses) (SSB 2003). Key enterprises have obviously become the backbone of the national economy, while they also play a leading role in local economies:

> In 2001 Guangdong province had 41 key enterprises, accounting for 0.2 per cent of the provincial total, but their sales revenue, profit and assets accounted for 18.5, 21.5, and 31 per cent respectively; Jilin province had 14 key enterprises, accounting for 0.5 per cent of the provincial total, but their sales revenue, profit and assets accounted for 43.3, 34.7 and 27.4 per cent respectively. Moreover, in 2001 China's key enterprises performed better than the Global 500 firms in terms of their 7.6 per cent of sales revenue annual increase while the latter decreased by 0.4 per cent, and of their 3.8 per cent higher proportion of profit revue.

> (SETC 2002)

In the coal industry, the positive effect of the government's industrial policy is reflected not only in Shenhua's case, but also in the entire industry. The industry turned a loss into a profit of 1.51 billion yuan for the coal sector in 2001, basically because of the profit contributions of 1.5 billion yuan from 25 key coal corporations. In 2002, 32 out of 33 key coal enterprises made a profit totalling 2.33 billion yuan (SETC 2003b).

Conclusion

In view of the dramatic impact of the global big business revolution, indigenous and competitive big business in China can only be built through powerful and effective government action. Even though China's industrial policy is still fragmented and has many errors, its positive effects on her economy and society can be clearly seen from the example of large enterprises in China in general, and Shenhua in particular.

However, there are many constraints to the implementation of China's industrial policy and the construction of large globally competitive corporations. Domestically, China has, for a century, lagged behind advanced economies in both technology and management. It is also a poor, regionally and sectorally unbalanced country, factors which have severely handicapped large firms in catching up with their international competitors. Since the reform in the 1980s, China has had to face three intertwined challenges from development, transition and globalization, which have deeply embarrassed the government. On the one hand, the government has demonstrated its ambition to build indigenous big business, but on the other it has to discount this ambition in order to save SOEs and TVEs. However, the implication is not that the government should give up its industrial policy, but that it should improve its effectiveness by dealing with the numerous conflicts of interest in a better manner.

Referring back to the beginning of this chapter on China's failure to build big business on a large scale, we may need to reconsider the criteria of 'success' and 'failure'. If we take a static comparison today between China's and global big business, then China's industrial policy could be judged a failure. But, in terms of the great progress that China's large firms have made, one might reach a different conclusion. Moreover, the constraints on improving China's industrial policy in the past were not only those special, internal difficulties that confronted China. At least as important is the fact that China's attempt to build large globally competitive firms coincided with the most revolutionary period in world business history, possibly even including the industrial revolution, as shown in the first section of this chapter. The period during which Japan and South Korea implemented their industrial policy to build global giant corporations was, in terms of the structure of the large capitalist firms, much less dynamic. This has presented a uniquely different challenge for China's industrial policy. Finally, the poor performance of China's industrial policy in the past does not imply that it is pointless to employ an industrial policy to catch up in the future,[15] although there are important lessons to learn about improving the design and implementation of industrial policy.

With regard to the improvement of the design and implementation of industrial policy, the first issue is how to select and to identify 'strategic' industries. Should the government give up those sectors that are strategically important, but which it may be incapable of building globally competitive large firms at present, such as aerospace? Should the government concentrate all its efforts on those industries which are strategically important and in which it may be feasible to build global giants such as the coal industry? These considerations suggest that the second issue is how to set different policies for different industries. In the case of China's coal industry, many areas require a well-designed industrial policy – from entry control, resource allocation, regulations to control rents and royalties on one hand, to the building of large corporations and the control of foreign investment on the other.

Figure 5.4 Building the globally competitive indigenous coal corporation Shenhua

Top left: Shenhua Group's headquarters in Beijing, making it more adaptable to market and customers. *Top right:* Shenhua's imported advanced equipment is extracting coal, contributing to Shenhua's world-level productivity. *Bottom left:* Shenhua has a dedicated railway and port facilities, saving over 40 yuan per ton in transport costs, and making it easier to reach customers. The photo shows the Shuohuang Line. *Bottom right:* Shenhua's success is attributed to many factors: high quality of coal, coalfield and coalmine, government support, excellent leadership and management, high technology and advanced equipment. This photo shows the high morale and team spirit among Shenhua staff and, in the background, the high quality of the coalmine.

6 Conclusion

Inter-relationship of the three challenges and the role of the state

A more successful transition away from a communist economy may, paradoxically, be easier to achieve with a strong state which is able to place the overall national interests above that of powerful vested interest groups. A self-reforming Communist Party may be the least bad vehicle available to accomplish this.

(Nolan 1994: 13)

First, even in mature market systems, government interventions are indispensable for organising efficient markets and remedying market failures. Second, improvement in human capabilities – the core of human development is an especially pressing responsibility of government in a developing country, such as China. Third, market institutions cannot be properly installed without the support of the state. Especially if China is to establish a 'socialist market economy', *the state has the obligation to mitigate the hardships caused by the market transition.* Finally, *as a giant developing country, China faces many challenges that cannot be settled by market forces alone.*

(UNDP 2000: 3, emphasis added)

Introduction

During the last few decades, there has been a major swing in attitudes to the nature and role of the state,[1] from market failure requiring government intervention to government failure inviting market liberalization again (see Killick 1989; Wade 1990; Chang 1996, 2002). The swing generated an increasing realization that there is no one kind of economic truth which holds the key to fruitful analysis of all economic problems, and the merits of both state and market should be recognized. Moreover, since the 1990s there has been a growing recognition of the role of the state, which is believed to play a more important role against the background of globalization and the growth of big business in this new millennium (Nolan 2001a, 2001b).

Many argue that the Chinese government intervenes too much in its economy (e.g. Sachs *et al.* 2000; Sachs and Woo 1997; World Bank 1995, 1996; Prybyla 1990). However, as shown in the case of the coal industry, the major problems resulted less from excessive intervention than from insufficient or misguided intervention. This final chapter aims, first, to examine the complex inter-relationship between the three challenges, which require involvement of a strong state; second, to summarize the successes and failures of the Chinese government in dealing with the three challenges in the past two decades; and finally to propose how the government should deal with the three challenges in the future.

The inter-relationship between the three challenges

The inter-relationship between the three challenges – from *development, transition* and *globalization* – is multi-faceted, with both positive and negative elements. The positive relationship suggests that the Chinese government should tackle the three challenges together so as to generate synergies between them; but the negative relationship requires that the government should exert control over the sequence and degree to which these three challenges are pursued or restrained.

Positive interaction among the three challenges

Economic development was the first priority when China started its reform in 1978. To serve this ultimate purpose, China started to reform its system and open its doors to the world. The following reforms greatly assisted the development of China's coal industry in particular and economic development in general.

The first obvious example was that, when coal shortage severely impeded economic development in the earlier reform period, the Chinese government transformed the system from one of SOE coalmines dominating the industry to one that encouraged the new entry of TVE coalmines to compete with SOE coalmines so as to force the latter to reform and strengthen themselves. The new capacity from those non-SOE coalmines together with the improved capacity of SOE coalmines finally removed the chronic coal shortage problem so as to push economic development into higher gear.

The second obvious example was that 98 per cent of Shenhua's equipment was imported, which is, without doubt, one of the most important determinants of the group's competitive capacity. Shenhua's senior officials recall that at the beginning of its construction in the early 1980s, the government did not realize that Shenhua[2] should be built as a globally competitive large coal corporation with modern equipment, but still envisaged it as a traditional labour-intensive and small-shaft coal company.[3] Numerous overseas visits and investigation of coalmines in America, Australia, Germany, Britain and South Africa, and comparison of productivity between imported and domestic equipment transformed the views of the senior officials.

As the world's leading coalmines were upgrading to use intensive use of high technology and high productivity levels, China's traditional small, labour-intensive coalmines would not be able to catch up with the world level. This would be a great disadvantage for China's coal industry in the future as China had decided to open its doors to the world.⁴ Finally, by 1989, the government had changed its view significantly and decided that Shenhua should be built as a large state-of-the-art modern coal corporation. Since then, Shenhua has been the first priority in the coal industry, supported by over $9.2 billion in investment, with exceptional import of advanced mining equipment and technology, and a reduced labour force. This example not only shows the importance of the open-door policy to China's development, but also reflects the pragmatic feature of her development path: 'groping for the stone to cross the river'.

The third example was that when coal was in surplus supply and when competition in the domestic market led to large losses, the government implemented a list of policies to encourage export of coal. Consequently coal export has grown dramatically during the last three years. China's coal has been recognized as a powerful rival for global coal giants. These benefits reinforced the argument that China should transform more quickly and integrate with the world more completely.

While transition and the open-door policy assisted China's economic development, the dramatic pace of that development during the last two decades has in turn allowed China to further its transition and to benefit more from the open-door policy. Examples in Chapter 4 showed that the transition of traditional SOEs, such as in Jixi, was made more acceptable by giving them sufficient time to realize the causes of their difficulties and to maintain basic living standards; the investigation in Chapter 5 showed that the building of large modern coal corporations has given China more advantage when they integrate more into the world economy.

This sort of positive relationship reinforces the view that the Chinese government should embrace the three challenges *transition, development, globalization* together to take advantage of all three.

Conflict among the three challenges

As shown above, transition and integrating with the world may assist development, but this mutual reinforcement might not happen automatically. By contrast, there are consistent and outstanding conflicts among the three.

The analysis in Chapter 3 showed how the explosive growth of TVE coalmines was a response to the development requirements of both the rural and the national economy. This development pressure initiated intense conflict between TVE and SOE coalmines. It was the development of TVE coalmines that first destroyed the monopoly profits of SOE coal bureaux, and later drove most SOE coal companies into financial difficulties, forcing them to restructure or reform themselves, or eventually be bankrupted. However, the SOEs' financial difficulties could not be solved without a higher level of development to make the transition affordable, requiring assistance for SOEs, from direct cash injection to innovative laws and regulatory support. Consequently TVE coalmines were forced to close to allow a gradual or a delayed transition for SOEs. It is not an exaggeration to say that the struggle between SOE and TVE coalmines is a matter of life and death (*nisi wohuo*). Obviously the entry of more TVE mines will lead to more bankruptcy of SOE mines, and keeping more SOE mines alive will force the government to close more TVE mines.

Meanwhile, inadequate integration with the world economy would not assist China's economic development. Those who encourage China to open the door widely usually believe that the huge foreign direct investment (FDI) it is likely to receive would greatly assist development. However, FDI and foreign technology do not enter those industries, sectors or products where they are most wanted by China, but come in based on what the foreign investors most want, such as highest profits or fast expansion of the market share. It is especially notable that while both coal and oil are energy industries and both are taken as strategic industries by China, FDI had particular interests in the oil industry, but almost entirely neglected the coal industry, as shown in Chapter 2.

In addition, the challenge from globalization appeared much stronger than before, as reflected in the global concentration of mining and the emergence of global mining giants as described in Chapter 5. Meanwhile, China's huge coal market has attracted almost all major global mining giants into China to establish branches or joint ventures. This will force China's coal industry to confront significant challenges directly. It is a big question whether China's small and weak, or large and weak coalmines are able to compete face-to-face with these global giants. It is also a big question as to whether China, due to the constraints from her low level of development and early stage of transition, has adequate capacity to build sufficient globally competitive coal corporations like Shenhua.

Although Shenhua represents the hope for the national coal industry from a long-term perspective, the messages that have emerged from the previous chapters are quite blunt. First, the government lacks financial capacity to build more 'Shenhuas' to compete with global giants, involving as it does huge investment in mine design, construction, equipment import and new infrastructure. China planned to spend $9.2 billion on Shenhua, compared to the total £3 billion in 1997 for the whole industry. This is the underlying reason why most investment in traditional SOE coalmines ceased following the concentration on large investment in Shenhua. Second, if the government is to assist large coal groups that are emerging from the restructuring of this industry, it needs to concentrate coal production in a much smaller number of large companies; consequently a large number of SOEs and TVEs need to be merged, closed or sold. Building 'Shenhuas' also implies that, by using more advanced and automated equipment, more employees will be replaced. Finally, if these SOEs and TVEs cannot be 'killed' or 'closed', Shenhua has to engage with them and even 'look after' them, even though it is expected to compete in the international market.

These conflicts of interest indicate that the degree of success in building Shenhua(s), and the extent of benefit from integration into the world economy, depend on China's resolution of other developmental issues, including whether TVEs could be closed; whether SOEs could be transformed; and whether the government has sufficient funds and administrative capacity to build enough Shenhuas. These constraints determined that the Chinese government had to be cautious and selective in its integration with the world, while deploying an industrial policy to protect its economic development.

Government dealt with three challenges: achievements and failures

Being aware of both positive and negative inter-relationships between development, transition and globalization, the Chinese government did play a significant role, deploying an industrial policy which, though not carefully designed, aimed to maximize the gain and minimize the conflict between the three challenges.[5] The major achievements and failures of the government in dealing with the three challenges in the case of China's coal industry have reflected the complexity of these challenges, the embarrassment for the government, and the consequent, sometimes unavoidable sense of failure.

Promoting the entry of TVE coalmines

The government encouraged the development of TVE mines by implementing policies of very low entry barriers, liberalized coal prices (while SOEs' coal price was still controlled), low tax and absolute management autonomy. The achievements of this policy were obviously reflected in the end of the coal shortage, the faster pace of economic development, and rural development.

However, the government failed to present a comprehensive long-term policy for the development of TVE coalmines. First, the central government had no policy for entry control during almost the entire period of the development of TVE mines. Second, there was also no centralized unitary organization to manage TVEs, with regard to their finance, employees, safety, resource allocation, access to railways, quality control, or to remove the local protection preventing cross-border mergers and cooperation. Finally, the central government implemented an at-a-stroke closure policy with no compensation. This is obviously an inappropriate policy for legal TVE mines, those areas which urgently need coal, but whose coal output is very limited, and those areas whose coal quality and geological condition are very suitable for extraction of TVE mines. It is safe to say that the TVE mines' huge problems were closely related to the almost complete absence of management from central government.

The problems for TVE mines are also fundamentally rooted in the long-running vagueness regarding their ownership, in which the government also failed to play an effective role. Most TVE mines are not purely privately owned, because even when there is a private proprietor, they only get the mandate to extract coal, not to own the coalfield. Meanwhile, TVEs are supposed to share local social obligations and their profits with the local community. They are also not purely collectively owned, because even in those nominally owned by local government, the coalmine is actually contracted out in different ways, which give the 'contracted owner' or manager almost absolute authority to run the mine and allocate the profits. While the non-private nature of TVEs leads to insecurity over their future so that a focus on short-term extraction is popular, their non-public nature enables them to maximize short-term profits. This is the basic reason why the development of these mines had so many negative consequences.

The government failed to issue any policy statement securing the position of TVE mines in the economy, as they had done with the policy on household responsibility, promising that that policy would last 50 years without change, so as to encourage peasants to operate their land with a long-term view. Moreover, the government did not implement any suitable reform policies, such as a shareholding system to clarify TVEs ownership, until the late 1990s.

Controlling the speed and sequence of transforming SOEs

As shown in Chapter 4, China did not privatize SOEs but transferred them step-by-step: from awarding more autonomy to firms, to forcing them to compete, to selling small ones, to reorganizing medium-sized and large ones, by bankrupting some and restructuring the remainder into large corporations. This strategy, first, safeguarded the consistently high rate of economic development by supplying sufficient energy sources; second, it mitigated SOE workers' suffering by gradual rather than dramatic restructuring while establishing a compatible social security system to smooth the transition pressure; and third, it enabled the government to take advantage of its ultimate unitary ownership to re-organize the unprivatized SOEs into large modern corporations to confront the challenges from globalization.

Meanwhile, there were numerous failures relating to the government's policy. The 'contract responsibility system' (CRS) caused a common problem of managers taking the short-term view. This sort of short-sightedness might not be significant in some sectors, but can do serious damage in the coal industry. The stable output and profitable operation of a coal company depends on a carefully designed long-term plan and close coordination between resources available, the speed and quality of extraction, workface preparation, and many other efforts. The CRS reform was constrained by having just three contracted targets for output, subsidy and infrastructure. For more effective performance it should have had more related targets including those for safety and preparation for new capacity. In Jixi and many other SOE coalmines, the loss of balance between resource, extraction and preparation in the 1980s and early 1990s (the contract period) has had a negative effect for many years, not only on lost current output due to no workfaces being available from which to extract coal, but also on future output, safety, and development due to historically low investment in resource and geological surveys, R&D, safety supervision, and equipment upgrading.

The dual-track price system was an innovative price reform, which helped to keep China's transition smooth and stable. It was important to control the price of coal from SOE mines to ensure that increases did not have a dramatic effect on other major industries or the economy as a whole. But there was no reason to hold the coal price so low for so long while liberalizing earlier and more quickly other equally important materials prices. Sectors which benefited from these relative price rises included electricity, freight and timber, which put the coal industry at a big disadvantage.

Moreover, the government improved other sectors' wage levels much more quickly than those of the coal sector, which in fact downgraded miners' standards of living. At the same time the government kept the controlled price of coal low, which prevented coal companies from making profits and from improving coal-workers' incomes.

Building globally competitive indigenous firms

During the development of Shenhua, the central government shifted its preference from traditional, small-sized, labour-intensive mines to large modern long-wall mines with a smaller workforce but more advanced equipment. This reflects the government's ambition to deploy industrial policy to build globally competitive indigenous coal corporations: providing large funds, allowing labour to be replaced by imported equipment, building dedicated railways and ports to maximize economies of scope, and developing a coal liquefaction programme using imported technology. Besides Shenhua, the government is intending to reorganize the thousands of coal companies currently in operation into seven large coal groups at national level.

However, the government failed to ensure that Shenhua avoided being constrained by the problems of traditional SOE and fast-growing TVE coalmines. The failure arose from Shenhua's acceptance of Five Western Bureaux, its unrelieved social welfare responsibilities, and its lack of autonomy making it unable to refuse unreasonable obligations imposed by central government. Shenhua is, despite all the changes, still not a modern corporation within a modern corporate system.

China's admission to the WTO makes the entry of foreign mining companies into China much easier than before, but the government has not reached a clear view on whether it should encourage foreign investment in the coal industry, on how to cooperate with foreign investors, and on taxation and royalties for foreign-owned coalmines.

As regards the proposed seven large coal corporations that the government intends to build, there have been no detailed proposals for how to reorganize China's tens of thousands of coal companies into these groups. An official in the former SETC conceded that 'we cannot force them [to merge], we just guide them'. But on how to guide, it still remains unclear.

Decentralizing administration in the coal industry

Chapters 3, 4 and 5 present three case studies from three categories of coal producers to illustrate each of the three challenges. If these coal producers are viewed as the hardware of the coal industry, then the government's administration, institutions, laws and regulations, and the coal transportation and sales network are the corresponding software of the industry. It is through this software and reforming this software that the government confronts the three challenges.

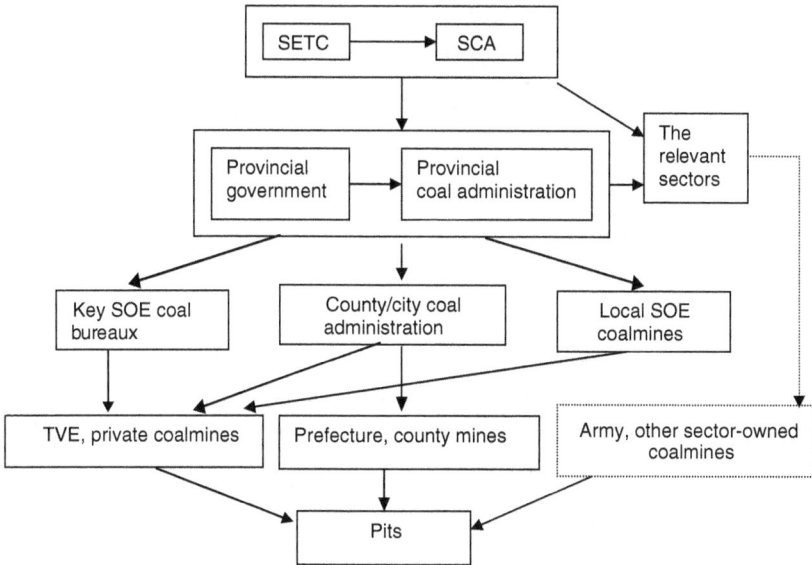

Figure 6.1 Administrative structure of China's coal industry since 1998

Sources: Interview source in China, 1999; 2000.

Notes
1   In 2001 the State Coal Administration (SCA) was also abolished and the State Coal Safety Prosecution/Supervision Administration was established at the same time by taking the functions of safety administration from former SCA, while China's National Coal Association was also established to be purely an association to communicate between government and coal companies.
2   Since the decentralization of 1998, sector-owned coalmines have also been released.

Despite the intention to build modern coal corporations through industrial policy, the administrative reform in the coal industry has been centred on decentralization and liberalization, aiming at separating government functions from direct enterprise management. It started by gradually awarding autonomy at local or enterprise level, and ended, with the abolition of the Ministry of the Coal Industry and decentralization of all key SOE coal bureaux to local provinces in 1998. The basic administrative structure of China's coal industry since 1998 is shown in Figure 6.1.

In addition to the investment reform, which changed the previous government distribution system to a bank loan system, the government's intention to withdraw its subsidies and implement the CRS reform were also part of the efforts to make SOE coalmines more competitive, and to separate administration from enterprises. With these reforms the coal industry has been transformed from its previous operation in the command economy system to a relatively competitive one. SOE coal companies have learned to be independent and market orientated. The government only holds the final decision-making power over the most important issues such as accepting foreign investment, selling any ownership rights, setting up joint ventures, applying for larger loans, becoming limited enterprises, entry to the stock market, and mergers and acquisition.

However, first of all, decentralization was not completed. The best example is where the central government forced Shenhua to 'merge' with Five Western Bureaux – the initial question raised by this study (see Introduction). The answer was found later during my visit to the Five Western Bureaux. The central government was indeed intending to decentralize the bureaux to Inner Mongolia. However, since most of these bureaux were large loss-makers and unlikely to survive, the Inner Mongolian government was not willing to accept them unless they were subsidized by the central government. However, the amount required by Inner Mongolia was too large to be accepted so that the central government had to force Shenhua to take them on.

Besides, the administration of the coal industry after decentralization seems not only structurally ambiguous but also weaker. The worst loss-making SOE coalmines still address their problems, such as problems related to the former Ministry of Coal Industry and investment in long-term uncompleted projects, directly to the central government rather than to their local provincial government. SOEs which were performing well could ignore the central government because they understood that the government would not invest in them in the near future.

More importantly, the central government becomes weaker due to lack of an alternative industrial policy to guide the industry after decentralization. Numerous important questions remain, including, how much investment should be made; how many SOEs should be bankrupt and how many TVEs should be closed; and whether and to what extent foreign investment should be encouraged.

When the government decentralized its administration in the coal industry it was obviously making an effort towards the system transition, which was in turn intending to promote development and globally competitive capacity. However, this transition effort was severely restrained by the government's capacity to establish alternative mechanisms such as industrial policy, and the low level of development as seen in the case of the Five Western Bureaux.

Improving the institutions, laws and regulations within limits

China's transition from a planned economy system to a market-orientated one requires its administration to set up, re-write or abolish numerous institutions, laws and regulations. Great achievements have been made in all these respects. A series of important laws and regulations have been issued, including 'Mining Resources Law of the People's Republic of China' (*zhonghua renmin gongheguo kuangchan ziyuan fa*) issued in 1986; 'Coal Law of the People's Republic of China' (*zhonghua renmin gongheguo meitan fa*) in 1996; and 'Administrative Regulations of TVE Coalmines' (*xiangzhen meikuang guanli tiaoli*) in 1994.

However, many institutions, laws and regulations remain unwritten, confused, or inadequate. China started to impose a 'resource tax' and 'resource utilizing supplement fee' (*ziyuan shui he kuangchan ziyuan buchang fei*) in 1994, which are similar to rent and royalties in a global context.[6] According to Chinese understanding, rent is designed to equalize the value of different grades of a resource, which leads to its generally accepted label of 'differential rent'. It is based primarily on the quantity and quality of the coal resource, cost of extraction, distance from customer, and land value. Royalties focus on adjusting the differences between rents of one country to the global market level. The purpose is to protect domestic coal resources by adjusting royalties to compensate for the displacement of domestic rent from the global market price (CERC 1995: 99–101).

However, due to the lack of a predetermined evaluation of coal resources at the national level, neither the resource tax nor the royalties incurred in China reflect the value difference between coals from different categories, reserves, locations, and infrastructure conditions. Nor do they reflect China's value level of coal resource compared with the global market level (see Table 3.6). In addition, quite a large number of coal producers, such as some illegal TVE coalmines, do not even pay the resource tax. Consequently there are two problems: first, about 4 billion tons of reserved coal resources are wasted each year in terms of the current low recovery level (SCA 1999a: 11) as there is no incentive for coal producers to maximize the recovery rate; second, China's coal producers are not actually competing with each other from an equal standpoint. Those occupying better resources are obviously in a more advantageous position.

This kind of situation might have been acceptable under the former planned economy system, because the government allocated resources for producers based on political considerations rather than economic efficiency, with prospering coalmines required to hand over their profits and loss-makers being subsidized by the government. However, this is obviously unreasonable in China's current market-orientated economy under which all coal producers have been asked to be responsible for their profits or losses. There is an urgent need to issue regulations to set different standards for royalty collection in different areas, resources and producers. This is key to determining the potential investment and profits for both domestic and foreign producers.

Because of the history of isolation in China's coal industry, there are insufficient laws, rules and regulations to assist multinationals in investing or setting up joint ventures. Pioneering multinationals complained that institutions associated with an overall planned economy environment were not compatible with private and foreign investments, and that the approval process was long, and complicated by vague interpretations of government policies and regulations.

An even more serious problem was that not only are laws and regulations insufficient, but also the available laws and regulations cannot be implemented in full or in a precise way. China is a vast, complex economy and society. Even in the Maoist period, a wide range of decisions was made at a local level. Moreover, the central government has intentionally streamlined the operations of the central bureaucracy, so that a great deal more detailed decision-making is carried out at lower levels than was the case a few years ago (Nolan 1999). This means that coal producers and customers must cope with policies and regulations issued by both central and local governments. For example, while tax collection is more unitary at the national level, fee charges vary and are not transparent at the local level (for more details see the next section).

These problems indeed troubled both domestic and foreign coal companies in China. However, laws and regulations are usually developed to meet the emerging commercial challenges, and therefore have to depend on the level of economic development.[7] When economic development is at an early stage and people are still struggling for survival, even though there are perfect laws available, it would still be difficult to compel people to follow them.

Failing to provide a sufficient and effective infrastructure for coal

The long distances between coal production areas and their markets, and the consequent long distances that coal has to be transported means that infrastructure bottlenecks are always a significant issue in China. In 2001 over 60 per cent of coal was carried by rail, and it accounted for 43 per cent of the total volume carried by rail.[8] The relationship between coal and transport is so close that the coal industry relies heavily on the development of the railway industry, and vice versa. In such a situation, whether a coal company can make profits largely depends on the quantity and quality of the rail service available for it. Even by the mid-1990s, rail capacity could meet only 70 per cent of demand for coal transport. The major causes of this problem were not only insufficient rail track, but also insufficient rolling stock, which could meet only 60 per cent, and at times, only 30 per cent of the total national demand. Just one province – Shanxi – lost 3.5 billion yuan each year because of insufficient rail capacity.

The government was aware of this high cost to coal producers and customers. For over two decades, from 1980 to 2004, several major dedicated coal rail lines – including the Shuohuang Line and the Baoshen Line – have been built. The Nankun Line (from Nannin in Guangxi Province to Kunming in Yunnan province), and the Jinjiu Line were also built to be partially dedicated to coal transport. In addition the construction of the new port at Huanghua, large coal ports such as Qinghuangdao, Rizhao and Tianjin ports have all been rebuilt to improve their coal-handling capacity. The improvements in infrastructure capacity dedicated to coal transport in the 1990s is shown in Table 6.1. At the same time, the government has organized a large amount of available freight capacity to transport coal, which has had over 42 per cent of total capacity for decades. In addition, the government organizes a nationwide Annual Coal Convention (*meitan dinhuohui*) to co-ordinate coal producers, railways and customers, to ensure that coal producers can gain access to the rail system and reach customers efficiently.

However, policy failures are still evident. First, the infrastructure for transporting coal is still very poor, and chronic bottlenecks still exist. The rail network has grown far more slowly than the national economy. It only increased by 4,200 kilometres between 1980 and 1997 (SSB 1998: 538). The average annual rate of growth of the rail network was just 0.9 per cent from 1980 to 1997, with only a marginally higher growth rate in the 1990s than in the 1980s. This compared with the average annual growth rate of 3.45 per cent for coal during 1978 and 2002.

Table 6 1 The coal-carrying capacity of rail and port systems 1990–2000 (Mt)

Rail coal-carrying capacity				Rail capacity to ports				Port capacity			
Line	1990	1995	2000	Line	1990	1995	2000	Port	1990	1995	2000
Daqing	33	80	100	Daqing	52	97	97	Qinghuangdao	74	79	119
Fensada	51	54	56					Tianjin			
Jinyuan	13	15	16	Jingqing				Huanghua			10
Shenshuo			35	Shuohuang							
Shitai	47	52	56								
Hanji	10	15	15	Jiaoji	5	15	20	Qingdao	5	15	20
Xinheyan	10	15	30	Yanshi	10	15	30	Rizhao	15	15	30
Longhai	11	19	29	Longhai	6	13	13	Lianyungang	12	13	13
Total	164	235	337	Total	73	120	160	Total	106	122	192

Source: MCI 1996: 126.

Notes

1 The relation between the three sections is that the first section 'Rail coal-carrying capacity' (closing coal producing area in inland China and not able to connect ports directly) is connected by the second section 'Rail capacity to ports' (able to connect ports), and is further connected by the third section 'Port capacity'. Lacking of any one of these capacities will lead to failure to deliver coal to some extent.

2 The three rail lines in the first section (Daqing, Fensada and Jingyuan) can be connected by either Daqing line or Jingqing line (with total capacity 52 Mt in 1990, 97 Mt in 1995 and 97 Mt in 2000), which can further be connected by either Qinghuangdao Port or Tianjin port.

3 Shuohuang in the second and Huanghua in the third sections have no figures because they had not been completed by 2000.

4 Shitai line in the first section is not able to connect any further capacity, which restricts its possibility to transport coal to port.

Table 6.2 The changes of capacity of various types of transport services 1995–2000

	Passengers (millions)		Passenger turnover (billion passenger km)		Goods(mt)		Goods turnover (billion ton km)	
	1995	2000	1995	2000	1995	2000	1995	2000
Total	11,726	14,586	900	1,206	12,348	13,257	3,573	4,220
Railway	1,027	1,016	355	439	1,659	1,655	1,287	1,335
Road	10,408	13,300	460	660	9,404	10,200	470	597
Waterway	239	200	17	11	1,132	1,200	1,755	2,220
Airlines	51	70	68	96	1	2	2	5
Oil/gas pipeline					153	200	59	62

Source: CASS 2001: 274.

Among various major transport systems, the development of the railway was the slowest (Table 6.2).

According to the Ministry of Railways, since 1997 service on the railways has been improved from 'bottlenecks' to 'basic compatibility', but not yet to 'complete compatibility'. However, the major contribution to this change relates to other types of freight transport services, where capacity improvements have been concentrated during the past two decades (Table 6.2). This has reduced some of the pressure on the rail services. Even so, both rail and port capacity is still insufficient. As the Ministry of Railway admits, these bottlenecks still exist during peak periods such as the Spring Festival and in the harvest season.

At the same time, almost all coal ports have very small berths and outdated coal-loading equipment which limits their capacity. Moreover, even the newly constructed coal port in Huanghua only has a design capacity of 30,000 tons capacity. It is said that is a result of the local provincial government insisting on building the port within the province even though local sea depths did not allow a large port to be constructed. This raises the question of why, first, central government did not block this local self interest, and second, if conditions were so limited, how Shenhua's chairman Ye Qing, once he had been appointed to Shenhua, was able to promptly expand the berth capacity to 60,000 tons before its completion. There was no doubt that central government failed to deal with local protectionism and ensure the largest possible port was built.

'Bottlenecks' resulted not just from the shortage of capacity, but also from the quality of the administration and service of the infrastructure. Chaotic arrangements for infrastructure access and unreasonable fees were the worst problems. Before 1993 allocation of coal from state-owned mines to users was tightly controlled as part of the mandatory planning system. Most coal from state-owned mines was purchased from state-designated suppliers. The Annual Coal Convention, usually hosted by the SPC and with coal producers, railway operators, and customers as participants, implements the state's General Allocation Plan (GAP). About a half of annual coal supplies are allocated according to letters of intent signed at the state-organized convention. A key function of the conference is to link coal allocation with transport facilities. Typically, sale contracts between producers and customers made at the convention have to match the state-allocated rail access quotas. The system is still in effect, although from 1994 the plan changed from being mandatory to having a guidance status. According to China Fuel Ltd, part of the State Power Corporation, 80 per cent of its coal was bought through the contract signed at the annual convention in 2000.

However, not all coal producers can participate in the convention and thus cannot get rail access quotas. Taking Shanxi province, the country's largest coal-producing province, as an example, only eight key SOE coal companies (Datong, PinShuo, Yangquan, Xishan, Fenxi, Luan, Jincheng, and Xuangang), and Shanxi Unified Allocated Coal Sale Company, Shanxi Coal Transport-Sale General Corporation[9] and the Taiyuan Coal and Gas Company could participate in the convention. Of these 11 participants, Datong and Pinshuo have their own connection to the railway network so they can load their coal using their own lines, and then access the national network by using their access quotas. Six other key SOEs could transport their coal through Shanxi Unified Allocated Coal Sale Company by using their quotas. Shanxi Coal Transport-Sale General Corporation allocates its rail access quotas to local SOE coalmines and its local transport-and-sale subsidiaries at city and prefecture levels, which then allocate to lower levels. Shanxi TVE Coal Transport-Sale Corporation cannot participate in the convention, so how much TVE coal can be transported depends on the quotas allocated to them from Shanxi Coal Transport-Sale General Corporation.

These disadvantages in access to transport led to local coal producers making much higher payments to coal transport and sale corporations. Taking a coal railhead (*jimeizhan*) at a key coal-producing city, Xiaoyi in Shanxi province, as an example, the selling price of coal at this railhead was 135 yuan per ton in 1997. However, the Prefecture Coal Transport-Sale Company charged several miscellaneous fees totalling 39.5 yuan per ton,[10] related fees of 15.5 yuan per ton[11] and a tax of 8 yuan per ton. Then, Xiaoyi City Coal Transport-Sale Company charged a loading and unloading fee, depletion fee and pollution remediation charge of 6.16 yuan per ton, tax of 3.18 yuan per ton, profit share of 8.26 yuan per ton. Finally, the coal producer received 53.8 yuan per ton. This example shows that transport and sale companies sometimes charged in excess of 60 per cent of the coal price, of which the fees accounted for 31 per cent of the price (Peng 1999: 80).

However, infrastructure problems did not end here. Once the coal gained reached the railway, the fees charged by different interchanges were equally unreasonably high and untransparent. With Shanxi coal still an example, those selling coal outside Shanxi needed to pay 39.80 yuan per ton as an 'out of province' fee. However, this fee was not imposed on key SOE's coal even when they sold outside Shanxi. Figure 6.2 takes a TVE coalmine HCC as an example to show that there were over 20 categories of fee payment totalling 170 yuan per ton for coal transported from Datong in Shanxi province to the port Tianjin in 2000. These fees excluded any black market deals to bribe the relevant officials to allow access to railheads and rail services.

Since coal producers and transport-and-sale companies endeavoured to gain access to railheads and railways, numerous powerful middlemen took advantage of coal

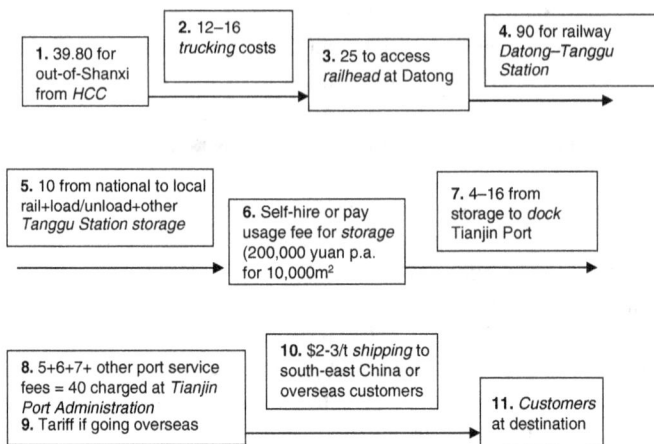

Figure 6.2 Costs incurred by coal producers between Datong and Tianjin port (yuan/ton)

Sources: Interview sources from the HCC; Datong Coal Administration; Tianjin Port Administration; and Department of Export of Shenhua Group, May 2000, June 2000 and September 2001.

Notes
1   The difference between costs for SOE coal and TVE coal was that SOE coal producers did not need to pay categories of 1, 2 and 3 as they had their own rail connection to the national rail network.
2   The production cost of coal for the HCC company, as a TVE coalmine, was 40 yuan/ton in 2000.

producers and customers. This not only made the coal market chaotic, but also encouraged corruption. Corruption occurred because many tried to own or operate railheads so that producers had to sell to them. It also occurred between railheads owned by railway operators and those owned by others and when railways followed the principle of 'white first, black later' – transporting high-value goods, such as white goods, first and black goods, such as coal, later. It also occurred when customers, to ensure that goods were transported punctually to meet delivery dates, had to pay 'preferential fees' (*dianzhuang fei*) to the railways, which rose from 5 yuan per ton in 1997 to 20–40 yuan per ton in 2001. During the last few years the situation was even worse because of the decision to speed up trains, which caused a severe shortage of rolling stock. Even key customers had to pay 'preferential fees' to get their coal transported earlier. Customers such as the State Power Corporation, which ordered about 210 Mt of coal each year through the annual convention, were guaranteed only 70 per cent movement by rail. It is no exaggeration to state that China's infrastructure service for coal was chaotic.

Clearly, these failures suggest that government intervention is much needed to provide sufficient and higher quality infrastructure services, to control the entry of coal transport-and-sales middlemen, to allocate rail access quotas in a more transparent way and to achieve better coordination between coal producers, transporters, sellers and customers.

However, it should be recognized that these failures related, in varying degrees, to China's low level of development and to specific features of a transition economy.

China could not build sufficient high quality railways because of limited financial capacity, technological under-development,[12] and also the nature of the transition economy.[13] When the country is still poor, railways are still state-owned, but operators are encouraged to be market-oriented, it would be very difficult to control self-interest activities such as railway companies operating railheads to make profits for themselves.[14] Even privatization, as in the UK, might not be the final solution to these problems. Although Britain privatized her national railways because of inefficiencies in the 1990s, the problems, such as insufficient investment and poor safety control, resulting from a privatized rail system even worse than before.

As analysed above, there were both achievements and failures when the Chinese government dealt with the three challenges of globalization, transition and development during the last two decades. However, if the two decades of economic development, social stability and national coherence are compared with other transitional economies' long-lasting economic downturn, social instability and even national partition, China's achievement was much more significant. The most important source of this success should be ascribed to the government's principle of putting the needs of development first, while controlling other challenges to serve this ultimate goal.

Government deals with the three challenges in the future

Even though the Chinese government was quite successful in dealing with the three challenges in the past, based on its development-first principle, it is still open to question whether this principle should be maintained or adjusted in the future, and it is still necessary to know how to avoid a recurrence of the failures analysed in the last section. The government's role needs to be renewed based on the changing challenges and their degree of significance for development in the future. For example, in the coal sector TVE mines were needed in the 1980s so as to promote the development of both the rural and the national economy, but now they are to be rationalized as rapidly as possible. An important recent example is in the financial sector. The government maintained state ownership of the financial sector and limited the entry of foreign banks and insurance companies, but now it is facing great pressures to liberalize the financial market completely as soon as possible.

Careful research is therefore a prerequisite for the government to deal with the challenges in the future, either at national level or in a particular industry. This section uses the example of China's coal industry to examine possible challenges in the future, then to recommend several key points which the government may need to take into account when dealing with the challenges.

Examining the challenges in the future coal industry

The findings in the last chapters suggest that the main challenges to China's coal industry in the near future are still from development, transition and globalization. The following sections will examine these three challenges respectively.

Examining the challenge from development

Although China's economy has been developed at high speed and people's living standard has been improved greatly, the country remains at a low level of development and still has massive poverty. In 2002 China's GDP reached 10,116 billion yuan, which was about $1,234 billion, ranking fifth in the world, but the average per capita GDP was $975 (CSD 2003: 16, 17), 129th in the world. Moreover, the absolute difference in per capita GDP between east and west China had been increased from 212.9 yuan in 1978 to 1738.5 yuan (at 1978 prices) in 2000 (CSSA 2002: 311). Consequently, China, especially in rural areas, is under irresistible pressure to reach a medium level of development by 2050. To achieve this goal, the economic annual growth rate has to be maintained at a certain rate.

Much more complicated than this, the 'Lewis model' of economic development with unlimited supplies of labour will still be a long-term fundamental factor in determining China's development strategy and produce. As Nolan (2003) concludes:

> China has a huge population, totalling almost 1.3 billion. The population growth rate is still quite fast, adding an extra 15–16 million people each year to the total. From 1990 to 1999, China's working-age population rose from 679 million to 829 million (SSB 2001: 97, 102), an increase of no less than 150 million in less than a decade. Almost 70 per cent of the Chinese population still lives in rural areas (SSB 2001: 95). Employment in the farm sector is stagnant (333 million in 1995, falling to 329 million in 1999) (SSB 2001: 372). It is estimated that there may be as many as 150 million 'surplus' farm workers (Yao 2002). This places a powerful constraint on the rate of growth of real incomes for low-skilled occupations in the non-farm sector (see below). … As the impact of the WTO increases, pressures on rural employment will intensify. The main alternative source of rural labour absorption, the 'township and village enterprises' (TVEs) stagnated in terms of employment creation (at around 127 million employees) after the mid-1990s.
>
> (Nolan 2003: 5)

This is the fundamental background factor in determining the importance and future direction of China's coal industry, and it indicates that the government should maintain its efforts to develop the coal industry so that coal's share of primary energy consumption will not be less than 50 per cent. In this way it will support the fifth largest economy in the world, and continue to promote rural development by absorbing surplus labour.

This is also the fundamental factor shaping future policy on the restructuring of SOE coalmines and the survival of TVE coalmines. It is, first, very important to keep and improve the productivity of major SOEs, as they still have the largest percentage of the capital, assets, equipment and technology in the coal industry. A certain proportion of TVE coalmines must also be retained, not only because SOE coalmines do not have the capacity to meet the total coal demand,[15] but also because the ability of TVEs to absorb surplus rural labour force must still be kept alive.

Examining the challenge from transition

Although transition involves a comprehensive system transformation for the whole society, the restructuring of SOEs must always stand at the centre. Losses and bad debts not only make SOEs suffer, but they also weaken the financial system and economic order, and even the reputation of reform. Examining the difficulties in transforming SOEs provides a key indicator of the challenge from the transition.

Attention must first be given to the financial provision that the government can afford for the costs of bankruptcy, redundancies and establishing an effective social security net. According to the most up-to-date internal documents from the CCC, to achieve the goal of closing and bankrupting 220 SOE coalmines and taking 80 per cent of SOE coalmines from loss into profit between 2001 and 2005, central government[16] must provide financial assistance of 17.5 billion yuan. This estimate was based on the fact that, by June 2000, there were 53 mines which had already been closed or put into bankruptcy at a cost of 4.27 billion yuan from central government finances (CCC 2000: 105).

The second question concerns the degree of toleration of redundancies and the possibility of social instability. There are many potential threats to China's social stability, including the huge population, the ever-larger social gap between rich and poor, the high levels of visible and invisible unemployment, and severe corruption. As shown in the coal industry, though social distress because of redundancies was the major cause of the frequent demonstrations, the anger over corruption was at least as important a cause. While it might be supposed that the hundreds of demonstrations at Jixi were due to people's financial difficulties, in fact most of them, including the first, were initiated by rumoured or actual corruption in their coalmines. Many demonstrators expressed their real concern in this way: 'if cadres stay with our workers, we are not scared of the difficulties. But now, cadres get richer and richer while our workers' suffering gets worse and worse. Moreover, their wealth comes precisely from the exploitation of our common property, therefore what they are enjoying is exactly what we are suffering' (Jinxi Mining Bureau pensioners, interviewed in Jixi city centre, 2000).

In China's coal sector even the most uneducated workers are able to describe two types of corrupt behaviour common in this sector. One is that SOE cadres made large private gains from lending, leasing or selling pieces of state-owned coalfield under their management to private persons. Another is that SOE cadres made private profits through the common triangle debt problems caused by *mo zhang. Mo zhang,* a term invented in the 1990s, involved paying bills with products or materials rather than with money. It was usually arranged between two or more companies paying bills by products, and without any amounts recorded in accounts; so the deal, including the value of a company's products, was decided by its present cadres in negotiation. It was possible for cadres to reduce the real value of their own company's commodities if they could get enough private payments from their counterpart. From 1993 to 1996 Jixi had over 1.3 billion yuan 'bills-on-the-way' (*zai tu mei kuan*), some of which were not paid, and some had been paid, but did not reach Jixi's bank account. By the end of 1999 when part of Jixi was proposed for bankruptcy, over 0.3 of 1.3 billion yuan 'bills-on-the-way' had been confirmed as 'bad debts', much of which was actually lost into private pockets.

This calls to mind the 'butterfly effect' popularized by chaos theory. China's history shows that in such an immense country, any small outbreak of anger and conflict could result in unforeseeable social chaos. The sources of potential conflict include wider social inequality between rich and poor, rural and urban, coastal and inland provinces. This is a major argument why China's transition should be cautious and gradual, but at the same time the government must take action to reduce these sources of conflict by attacking corruption more effectively.

Examining the challenge from globalization

The most important examination must be given to the possibility of foreign mining companies entering the market, how competitive their coal will be if the coal is to be produced in China, and how capable Chinese companies are to compete with them.

As seen before, while global mining giants have been strengthened through concentration, China's coal industry suffers from excess competition and lack of economies of scale. The enlarged gap between global mining companies and China's counterparts makes the future for most loss-making SOE coalmines and backward TVE coalmines bleak. Although foreign companies are not necessarily more competitive than Chinese companies when incurring different royalty and tax rates to them, they have no obligation to pay social welfare as domestic companies have.

A deep concern in China's coal community is therefore that the tens of thousands of domestic coal companies would be beaten by multinationals and that millions of coal employees would lose their jobs if multinationals were allowed uncontrolled entry. As officials openly admit, China's productivity and technology in the coal industry (including equipment construction) are at least 20 years behind the world's leaders

Another important examination must be made of both international pressure and China's ability to reduce coal production and consumption. China is one of the major targets for criticism by environmentalists and some world organizations. Together with pressure from inside China, the ratio of coal in the total primary energy consumption structure may have to be reduced from the current 68 per cent to 50 per cent by 2050.

However, for China the availability of oil and gas is much less than coal and they are also much more expensive. It is well known that China is constructing a huge west–east gas pipeline, which will stretch 4,200 kilometres and cost $18 billion. 'The government plans to double gas usage, to an annual 80 billion cubic metres, by 2005, and eventually use it to cut urban pollution by phasing out coal…' (*Business Week Online* 29 April 2002). However, at current prices, the clean-burning fuel can cost twice as much as average-quality coal. It is said that probably only customers in wealthy cities such as Shanghai and Beijing can afford gas. As a director of Shanghai's largest gas distributor company has said, if the industrial users can't afford the price, there will be no market for this gas. To use more gas is the future trend and is also ideal for the environment, which fact is well known, but obviously this would depend on the level of development in China.

Dealing with the three challenges comprehensively

The simple examination above can shed some light on government's role in dealing with the three challenges in this sector in the future.

Development will still be the supreme challenge

Development will still be a paramount challenge. The implication for TVE coalmines is that the government should develop a determined long-term policy on the development of TVE coalmines. Because of specific features of the coal industry and China's chronic transportation bottlenecks, the existence of small mines makes economic sense as they can effectively recover coal resources that could not otherwise be recovered by large coalmines. They are best placed to serve local customers. Moreover, without the contribution by TVE coalmines China may return to a coal shortage. Since 2002 there have frequently been reports of coal shortage in China.[17] The world's steel industry has even been severely affected due to China's short supply of coke, which in turn partly results from the large-scale closure of small coalmines.

The existence of TVE mines will also retain importance other than in a purely economic sense. Besides economic growth, successful development entails the stability of the vast population, the reduction of surplus labour force, the reduction of rural poverty and the narrowing of the gap between rich and poor. The pressure from rural poverty and the surplus labour force on economic development are observed nowhere more dramatically than in the coal industry. The coal industry provides extreme ease of entry for small-scale producers and they have made significant contributions to China's coal industry, to rural development and to urbanization.

Now that it has been shown that China still needs TVE mines, the government's major task should be to close unnecessary mines and improve the necessary ones. It was also suggested that a different closure policy was needed in different areas and for different categories of coal producers. For example, it might be appropriate to issue more licences to TVE mines in provinces with high coal demand (such as eastern China), and fewer in low-demand but high-supply provinces (such as western areas). It was justifiable to close mines producing coal with high sulphur and ash content, but it might not be correct to close those producing high-quality, environment-friendly coal.

More importantly, the government should strictly control the establishment and the quality of TVE coalmines, although it may be difficult to do so. In addition to other methods – such as taxation, royalties, access to transport, centralization of the issuing of licenses, establishing minimum requirements for capital, size, equipment, and health and safety to operate a coalmine – the most useful would be in controlling the number of TVE mines within a reasonable range. TVE mines should also be encouraged to strengthen themselves by mergers and acquisitions both within the TVE sector and with SOE mines or modern corporations. Shenhua's leasing of TVE mines to provide a flexible amount and quality of blending coal for export is a successful example.

Transition needs to be tightly controlled but well-designed

Transition needs to be tightly controlled because its challenge is going to become more intense. In the coal industry, there are two urgent challenges from transition: one is to deal with the financial problems of SOE coalmines, and the other is to create a controlled competitive market. This is not a completely free market, but a more fairly competitive market achieved by deploying a range of government administrative methods.

Before confronting these two challenges the government must have a comprehensive understanding of the coal industry and its relative importance, i.e. the proper position for the industry in the context of national development. Since the 15th CCP Congress the general strategy for SOEs was to 'enter some while exiting others' (*you jin you tui*), which means government withdrawal from some competing industries while concentrating on some natural monopolies such as the defence industry or strategically important industries such as energy. The information from the whole of this book implies that the government should considerably strengthen the coal industry because of its strategic importance. The large reduction in investment after the end of the period of coal shortage may prove to be short-sighted, because the industry not only needs to meet demand, but also needs to compete with other industries and multinational coal giants. The latter would not be possible with wildly uncontrolled small-scale coalmines. With this consideration firmly in mind, the government needs to work on the following.

First it needs to balance investment among various industries on one hand and within the energy industry on the other, and stop the trend of distributing less and less coal investment among them (see Figure 2.3). Increasing investment in coal could enable SOE companies to update their facilities and technology, to improve coal quality and after-sales service, to assist them in reducing overstaffing by providing subsidies or by using them more efficiently through increased investment. All of these measures are vital for improving competitive capacity of SOEs and ability to attract mergers during the cross-sector reorganization analysed below.

Second, the government could, as the result of having the proper position of coal in mind, consider encouraging or even enforcing large-scale reorganization of better resources and assets within the coal industry, and with other industries, such as power plants and railway or port operations. The prominent motive for this is the reduction of the huge transaction costs that currently exist. Meanwhile, reorganization could also solve SOE coal companies' financial difficulties by pursuing economies of scope. Traditionally China's industry as a whole was fragmented and segmented between ministries, each ministry often blocked the activities of other ministries. The coal industry has for decades suffered from not being able to explore profit-making potential from its high-value-added business activities, such as coke, electricity, chemical products, and construction materials, which were managed by other ministries. At present, other ministries covet the coal liquefaction business currently conducted by the Shenhua Group. The reorganization could assist in the emergence of large corporations capable of building their new portfolios by combining businesses from both low- and high-value chains of coal. The wave of coal-electricity 'marriages' (business cooperation), which has been a hot issue since 2003, originates from the reorganization of the power industry; this implies that companies from both the coal and power sectors will be even more active in pursuing such cooperation if coal companies become more attractive. Moreover, the reorganization of better assets into good-quality coal corporations would be vital if they are to get stock market listing so as to mobilize social funds. More importantly, the threat of the growing concentration in the global mining industry requires the Chinese government to deploy the strategy of strong-with-strong mergers (*qiang qiang lianhe*) rather than liberalize companies so that they become too weak to compete with the multinationals.

Third, during such a reorganization hopelessly unproductive coal companies may have to be transferred to other businesses or even allowed to go bankrupt. This will depend on available government funds and the degree of impact on coal supply and social stability.

Fourth, there is no reason to be optimistic about the ability of most SOE coal companies to be reorganized with assets from other sectors. The reason is that not many of them are attractive. This is determined by the fact that the coal industry still needs to 'work from the beginning' and strengthen itself. At this point, to create a controlled competitive market is vital both for the SOEs and other coal producers.

As the current coal market is so chaotic, to create a fair government-controlled competitive market will require a sequence of inter-related policies and detailed reforms. These include:

1 The current natural resource policy must be reformed. Coal is an important non-renewable natural resource, whose value changes with time, category, quality, utility, degree of availability, depletion, recovery rate of exploration, and location. Therefore, how to maximize its value and utility is vital for a nation's long-term sustainable development. In addition, the competition between different coal producers must be unfair if the value of their occupied coal resources is different but their payment in tax and fees shows no significant difference. Due to these natural features, coal should be extracted by precise payment, which should vary depending on its comprehensive value (not only the current value but the value in the future). Therefore, the current resource tax and royalty collection system should be changed from one based on turnover to one based on the value of coal exploited. The prerequisite preparation for this policy change is to have a comprehensive nationwide coal resource evaluation.

2   Based on the different advantages of TVE and SOE mines, it might be correct to allocate different coal resources to different producers so that the resource can be extracted at a maximized economic level by ensuring a suitable capacity of exploration.

3   It might be necessary to set different policies to different coal producers. Sustainable market competition requires fair competition. However, coal is more like a 'public good', which results in market failure. For example, the coal-producing cost of an old coalmine is usually higher than that of a new one because when the coal shaft becomes deeper, more advanced equipment and a greater labour force may be required. Again, producers who have better quality coal, or who have a shorter distance to coal customers have an advantage over others. Without government policy adjustment, those old coalmines, those in a poor infrastructure environment, or those in remote areas would be disadvantaged in market competition. This requires the government to set different conditions for different players including SOEs, TVEs, large coal corporations and probably the future joint ventures and foreign mining companies, by charging different prices for coal resources exploited, and incurring different levels of tax, royalty, and subsidies based on their history and natural condition.

4   As regards the excessive competition in the coal market which disadvantages coal producers but benefits middlemen, there is merit in trying to allocate the coal market at national level among different groups of competitors. The basic idea might be to set target customers for different groups of coal producers. For example, for Shenhua, the target should be the most important domestic customers and export market; for TVE coalmines, the target is local small customers and residential customers. In addition, market ranges could be set for different coal producers, with several competitors in each range so that coal producers can compete in a certain range of the market. The current Tenth Five-year Plan proposed to establish seven large coal groups by reorganizing the nation's hundreds of large coal companies. This reflects such market allocation idea, but the seven large groups may need to be allowed to compete in more than one sector of the market, otherwise monopoly in their allocated range of the market will be a problem.

5   To build an effective and efficient infrastructure system for coal is another important task for he government. This includes the coordination between coal supply, customers and transport to ensure that coal can be transported in time; reorganizing thousands of coal sale dealers; prohibiting the rail service department from operating railheads and other coal-related businesses; setting the measurement of fee charges to ensure its transparency; and coordinating different energy sources to supply the same areas.

Entry control on foreign companies and promoting the CCT

In principle, China should encourage foreign investment in the coal industry, which is helpful for renewing facilities and improving technology, for payment of bankruptcy costs, for increasing the coal-washing ratio, for building more pithead power plants, for development of the coal liquefaction and gasification projects, for the research and development of China's CCT, and for the cost of adopting CCT such as replacing old-fashioned industrial boilers.

However, although China has entered the WTO, there is every reason for the government to control the direct entry of foreign mining companies into the domestic coal market, from the viewpoint of energy security, for protecting the tens of thousands of weak SOE and TVE coalmines, and for preserving the potential profit from exporting coal for the domestic coal companies. Energy security is one of the most important considerations.

Coal for other countries may be just a fuel, but it is vital for the security of China's energy. As emphasized in the rest of the book, China's capability to substitute other sources of energy for coal is very limited.

Responding to the insecurity arising from as much as 64 per cent of imported oil currently having to pass through the Strait of Malacca, before 11 September 2001 China and Russia had reached agreement to lay pipelines to increase China's oil supply from Central Asia. However, immediately after the attacks on the US, China realized that it has to reconsider its strategy of oil security, and that it must adhere to a strategy for multi-channel supply of oil. 'The terrorist attacks on 11 September have objectively provided a pretext for the US to enter Central Asia in a way that complicates a once simple situation. This will be of far-reaching significance to the strategy for oil supply in China' (People's Daily Online, 26 September 2001). One of the important strategy shifts for China is to speed up its coal liquefaction programme, in addition to the previous policy of substituting coal for oil. It is based on such considerations that the government supported the Shenhua Group to start building a coal liquefaction plant despite both the financial and technological risks. The implication is clear, that China's energy security must still take coal as its central focus.

Conclusion: understanding China's development strategy

This research had the particular purpose to examine the question of how to explain China's success and failure, and how to understanding its development strategy. Based on an in-depth case study of the coal industry, this research has found the underlying answers in three key contemporary trends.

China has been experiencing the interlinked challenges from development, transition and globalization. The three reinforce one another while also disrupting one another. For example transition and integration into the world may assist development, but dramatic transition and liberalization of the domestic market to external market forces may impede or even reverse economic development. China has succeeded because it understood that development is a paramount challenge and an ultimate goal, so transition and integrating with the world must be taken cautiously to secure, rather than radically to disturb, economic development. However, as all three challenges are so overwhelming, to pursue one may have to delay or even sacrifice another. The best solution requires a complex mix of considerations in respect to these three challenges. This is why China has enjoyed qualified success, succeeding in some areas and failing in others. China's development strategy, though with some problems, has proved practical and effective in the often difficult conditions created by intertwined forces of development, transition and globalization pressures.

It is too simple to criticize the Chinese government for deploying gradualism during the last two decades. China has suffered a mix of the three greatest challenges, which have not been suffered before. She has handled them in an innovative way. That is, to take development as top priority, transform its system gradually and integrate with the world while ensuring that this promotes development. Gradualism was deployed not because it was without costs, but because it was suitable for China's complicated situation and appropriate to deal with the three challenges.

Socialism was unsuccessful in the race with capitalism in terms of promoting economic development. Once this was recognized, the former socialist countries were recommended to pursue a new development strategy by following the path of capitalism (such as by adopting 'people's capitalism' and free trade), but this has failed too. Looking back to those early capitalist countries, which were initially successful in their early development path, but now with the USA in the lead they are not only suffering a new cycle of recession, but also are increasingly assuming the characteristics analysed by Marx in the Communist Manifesto and other writings (Lefeber 2000: 542). The latter is suggested by frequent anti-capitalism and anti-globalization demonstrations, and even regional and international wars.

The failures and the conflicts, whether in former socialist countries or current capitalist countries or between them, are deeply rooted in those consistent common issues for the human being: underdevelopment or unbalanced development, inequality and injustice, greed and individualism, and lack of necessary understanding and tolerance for each other and effective international cooperation. Unfortunately, these were precisely the basic causes for the recent two world wars and numerous regional wars. It is the time for both developed and developing countries to rethink their own development paths and to learn to respect other development paths chosen by other countries.

China's experience provided a great contribution to this re-thinking process. That is, while it seems impossible to compromise on the full development agenda for different nations, societies, cultures and religions, it is reasonable and also important for different countries to choose different development paths based on their specific features. This is exactly as Dobb predicted as early as 1967: there are probably no identical historical objectives for different systems.

Dobb illustrated the development paths of capitalism and planned economies as shown by lines A–B, B–C in Figure 6.3. While describing the development path of early capitalism as the cycled master's route AB, Dobb gave credit to the planned economy which might be able to act as a 'rational dog' to follow his master by a shortest straight route CB, rather than the curve CB. However, Dobb (1967: 88) made an impressive conclusion which can precisely illustrate the theme of this book:

This analogy, however, is limited. It assumes, for one thing, that both economies pursue ... the same historical objective. As we have seen, *they*

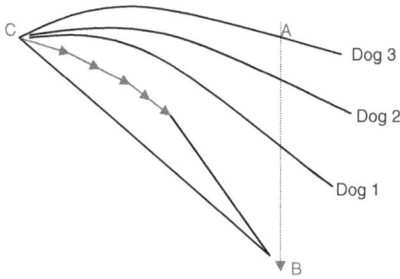

Figure 6.3 The pursuing path of the dogs[18]

Note
Lines A–B, B–C were illustrated by Dobb (1967). The paths for dogs 1–3 are added to illustrate the theme of this thesis.

> *may do no such thing; and it might be more proper to say that the crucial*
> *difference between the development-path of these two types of economy is*
> *that the historical objective which it is practicable for one system to set itself*
> *is quite different from that which the other system can have … on its agenda.*
> (Dobb 1967: 88, emphasis added)

China's experience shows that Dobb's general perspective on the development paths in different systems is still valid. There are many 'national dogs' that are searching for their development paths, but it is not necessary that they follow early capitalism's path AB to pursue the destination – point B. The pursuing paths of the dogs are uncertain, and the speed and distance are also uncertain. If China's development path is illustrated in the pursuing path of dog 1, dogs 2 and 3 refer to the development paths of any other single country; they have the right to choose their own path and destination of development. The judges can assess the race, the coach can train the dogs, but nobody can replace the dogs in the race, so everything depends on the different dogs' energy, wisdom, caution, passions and all sorts of prevailing conditions. We still don't know where and when the Chinese dog can reach the destination. An equally important question is, where is the destination?

Postscript

At a seminar of the Chinese Big Business Program there was an intense debate on China's development path between two scholars.

Doctor A: I support China's development strategy of 'groping for stones to cross the river'. It may be the second best, but 'the best is the enemy of good', and at least it is 'good'.

Doctor B: If it is only the second best, where is the first best strategy?

Doctor A: I don't know.

Doctor B: So, the second best is really the best?

Doctor A: Unfortunately, you are right. The second best is really the best.

Figure 6.4 How will the government deal with the three intertwined challenges?

Top left: the former Ministry of Coal Industry building, the Ministry was abolished in 1998. After decentralization the administration of the coal industry seems weaker. *Top right:* Xiaohenshan Mine of Jixi Mining Bureau contracted out this piece of coalfield to a private mine company. While the Xiaohenshan was bankrupted in 2000, the latter is still prospering, by hiring redundant workers and 'borrowing' facilities from the former. *Bottom left:* to 'look after' loss-making SOEs Shenhua contracted this project to Jixi's workers. *Bottom right:* to deal with the TVE problem, Shenhua leased this small TVE coalmine in Shanxi province as a blending coal base. These photos reflect my deep worry for the Chinese coal industry: how is this weaker administrative structure going to deal with these three intertwined challenges in the future?

Notes

Introduction

1 CBEs were renamed in March 1984 when the government issued the agreement to the Ministry of Agriculture with 'The report of generating a new situation for commune and brigade enterprises' (Ye and Zhang 1998: 33).
2 This might be another reason why transition should be cautious, because there is no existing ready prescription.

1 The challenges facing China

1 Under the First Five-Year Plan (1953–7) agriculture was to receive only 7.6 per cent of the budget compared with 55.8 per cent received by capital goods industries and 19.2 by transportation and communications (Breth 1977: 4). The per capita grain output in 1978 was the same as it was in the mid-1950s.
2 In theory, Mao was highly confident that 'the absurd argument of Western bourgeois economists like Malthus that increases in food cannot keep pace with increases in population was … thoroughly refuted in theory'. He believes '[i]t is a very good thing that China has a big population. Even if China's population multiplies many times, she is fully capable of finding a solution; the solution is production' (quoted from Smil 1993: 14). In practice, Mao ignored those who criticised his population policy such as Ma Yinchu, and rewarded those 'glorious' mothers who had more children.
3 Joan Robinson (1936) invented the concept 'disguised unemployment' to describe those workers in developed countries who accepted inferior occupations as a result of being laid off from industries suffering from a lack of effective demand. Later, Buck (1937), Warriner (1939, 1948, 1955) and Rosenstein-Rodan (1943) found that a large percentage of agricultural labour was idle for substantial periods of the year in less developed countries such as China as well.
4 Lewis claims that, 'if the capitalist sector produces no food, its expansion increases the demand for food, raises the price of food in terms of capitalist products, and so reduces profits. This is one of the senses in which industrialization is dependent upon agricultural improvement; it is not profitable to produce a growing volume of manufactures unless agricultural production is growing simultaneously' (Lewis 1954: 141).
5 Except Romania and Bulgaria.
6 They conclude that the short-term costs of trade liberalization for goods examined in their study will be substantial in terms of lost domestic output and lost jobs. The long-term benefits, however, would provide some $35 billion worth of consumer benefits.

7 This was recognized as a major cause of the Latin American financial crisis in the 1980s and in 2002 (Singh 2002) and of the 1997 Asian financial crisis (Stiglitz 2000).

8 'The bourgeoisie ... by the rapid improvement of all instruments of production, by the immediately facilitated means of communications, draws all, even the most barbarian, nations into civilization. The cheap prices of its commodities are the heavy artillery with which it battles down all Chinese walls, with which it forces the barbarians' intensely obstinate hatred of foreigners to capitulate. It compels all nations, on pain of extinction, to adopt the bourgeoisie mode of production; it compels them to introduce what it calls civilization into their midst, i.e., to become bourgeois themselves. In one word, it creates a world after its own image' (Marx and Engels 1888: 84).

9 'In the past the organization of trade had been military and warlike; it was an adjunct of the pirate, the rover, the armed caravan, the hunter and trapper, the sword-bearing explorers, the planters and conquistadors, the manhunters and slave traders, the colonial armies of the chartered companies. Now all this was forgotten' (Polanyi 1957: 16).

10 Griffin has more details on these points. He lists three consequences that globalization has brought and developing countries must be aware of. First, it has created opportunities for material betterment. The benefits include the creation of new markets, the transfer of technology, the attraction of more foreign investment and so on. Second, however, globalization has unleashed economic forces which tend to increase inequality in the global distribution of income. One reason for this is that the reduction of trade restrictions has been uneven. Another is that capital flow tends to be attracted to the already industrialized and to the rapidly industrializing countries, where the returns on investment are relatively high, and to by-pass the poorest countries, where profit rates tend to be relatively low. A third consequence is that it has weakened the ability of individual states to manage their economies (Griffin 1996: 126–7).

11 'Even with greater willingness to turn to foreign sources to finance its modernization and development process, China's policy toward external borrowing has been relatively conservative. Borrowing has remained predominately medium- and long-term from official sources, and a substantial proportion of it has been on concessional terms ... [China's] prudent debt management and an improved monitoring capability have served to limit the ratio of total debt outstanding to GDP to about 16 per cent in 1992, with about 20 per cent of total debt being on concessional terms. Throughout the reform period, China's debt–service ratio has remained much lower than in most other developing countries' (Bell *et al.* 1993: 9).

12 Such as transparent market regulations.

13 The four basic principles are: to hold on to socialism; to hold on to dictatorship of the proletariat; to hold on to the CCP leadership; and to hold on to Marxism-Leninism and Mao Zedong's ideology.

14 This is exactly what Huntington (1968) stated: lower level of political participation combined with lower political institutionalization will not lead to political disorder.

2 The challenges facing China's coal industry

1 During this period coal had first of all to serve the needs of the steel and defence industries. The output goal was 113 Mt by 1957, up from 32 Mt in 1949.

2 In response to this, in October 1958 the Ministry of Coal Industry announced that 'All the people exploit coal' and 'Taking scattered small mines to deal with the scattered small steel mills' (CCCIEC 1989: 44).

3 The coal industry prepared vigorously for the war by constructing mines behind the frontiers (*hou fang*). Meanwhile, to change the situation of transporting coal from

north to south, coal was required to develop the area south of the Yangtzi River. All this meant that the coal industry had to shift its focus to new locations.

4 There are even more estimations which believe China will consume more than this. For example, the US Department of Energy (DOE 1996, pI-2) estimated China would consume 1,829 Mtce by 2006 and 2,969 Mtce by 2015.

5 China's current financial capacity cannot afford the import of large amounts of oil, or the cost of replacing current inefficient and polluting coal-fired equipment.

6 Because demand for coal has increased and also because a large number of small coalmines have been closed.

7 This was in December 1981 when the State Council agreed to the National Energy Commission's proposal, namely 'Request for instructions on the subject of various problems of economic policy with regard to locally run state coalmines' (CCCIEC 1989: 105).

8 It was decreased from 23.7 Mt in 1995 to 19.95 Mt in 2000 due to a sulphur control policy, and the use of high-sulphur coal was prohibited (State Environment Protection Bureau 2001).

9 The regulations governing the emission of sulphur dioxide and carbon dioxide have the following objectives:

1 to control sulphur dioxide emissions resulting from the burning of coal by a) limiting the exploitation of high-sulphur coal, b) increasing the amount of washing and separating of kinetic coals, c) limiting the sulphur levels in coals burnt in urban areas;

2 to apply strict controls to sulphur dioxide discharges from a) thermal power plants, b) industrial boilers and c) industrial ovens and combustors;

3 to step up the construction of urban energy infrastructures and control the amount of pollutant discharges from households;

4 to control the discharges during the process of coal treatment at every stage (State Environment Protection Bureau 2001).

10 When, in 1998, China washed 310 million tons out of a total 1,232 million tons, the wash rate was only 25 per cent. By 1998 there were 1,527 washing plants with a capacity 494 million tons, of which key SOEs had 233 plants capable of washing 347 million tons, while most TVEs did not have a coal-washing facility, although they produced about 50 per cent of the coal.

11 It was announced that 136 small power plants with a capacity of 1.8Mkw were closed in 1999 (CEYEC 2000: 268).

12 As will be shown later, the high cost of transporting coal by rail may also be reduced, and the huge pressure on rail access to coal transport can also be partly removed. According to the new policy, new or planned large coal companies are required to build their own pithead power plants. As far as we know, there are more than ten large and medium-sized pithead power plants that are planned to go into operation in the near future. The recently established coal corporations (such as Shenhua, Yanzhou, Zhungeer, and Xishan Coal-Electricity Limited Corporation) have all built their dedicated pithead power plants.

13 It was first set in the First Five-year Plan (1953–7) when coal from 76 key SOEs was set to be unitarily distributed by the state, and nothing could be changed in this category without the approval of the SPC. By 1985 410 Mt of coal was still unitarily distributed by the state (CCCIEC 1989: 561), compared with the 406 Mt by the key SOE bureaux and 872 Mt by all other producers in that year. Except for a small proportion of the output of local state-owned coalmines, this 410 Mt was mainly from key SOE coalmines. This implies that almost all the output from key SOEs was distributed by the state.

14 From 1949 to 1951, the Ministry of Fuel Industry set up seven sale departments around China to be responsible for distribution or sale of SOE coal to planned customers (CCCIEC 1989: 561). This system was stopped in 1952, but reactivated

from April 1978 to November 1981. To ensure raw material supply for key projects, the State General Bureau for Raw Material was established in 1978 to centralize the management of all raw materials, including coal. The Ministry of Coal Industry was in fact in charge of production only, while State General Bureau for Raw Materials was responsible for transportation and sales (CCCIEC 1989: 562).

15 One leg was the capacity of SOE coalmines and the other was that of TVE coalmines.
16 During the Ninth Five-year Plan 1996–2000 the total FDI used in the coal industry was $1.14 billion. Of which about $100 million was used to develop coalbed methane, and $480 million to import mining equipment from advanced countries (SETC 2002, The Tenth Five-year Plan of the Coal Industry).
17 This used to be a joint venture between Occidental Petroleum and the Pingshuo Coal Industry Company. The mine opened in 1987 'after years of planning and negotiations'. At that point it was the largest single joint venture project in China. The venture encountered many problems, from recruiting the necessary Chinese personnel to obtaining necessary materials, such as winter fuel. However, the major problem was the collapse in the international coal price at that time.
18 China was considered by international mining organizations to be one of the countries with the worst investment environment for mine exploration and, because of this, China did not win any of the 116 large non-oil and gas mining product development projects which were the recipients of global investment by multinationals in 1999 (Zhu 1999).
19 Most of China's coal has high natural levels of sulphur of around one per cent and ash levels up to 20 per cent. Over 80 per cent of coal used is unwashed.

3 Development: the rise and fall of TVE coalmines

1   Taking 1999 as an example, the national total number of TVEs was 20.7 million with 127 million employees, of which collective TVEs numbered 0.94 million with 43.69 million employees, private enterprises were 2 million with 28.51 million employees, and individual enterprises were 17.69 million with 54.84 million employees (CTVEYEC 2000: 114).
2   Shanxi province alone had 3,616 small private coalmines.
3   Those numbered 51,000 collectively owned and 12,000 privately owned.
4   Of more than 30 towns in Shenmu, the revenue of some of them came almost completely from TVE coalmines. The town of Daliuta is located at the centre of the famous Shenfu coalfield of the Shenhua Group. Daliuta has 376 km^2 of land and a population of 12,000. Government revenue in 1999 was 6.3 million yuan, of which 5.9 million yuan or 94 per cent of total revenue came from TVE coalmines directly, excluding contributions from coal-related income. There were 48 TVE coalmines producing 1.8 Mt of coal.
5   Baode county was also rich in coal and TVE mines were also developed.
6   For example in 2000, transporting coal via the DaQing Line cost 150 yuan/ton, via BaoShen–DaQing cost 170 yuan/ton, via the Shenhuang Line cost 120 yuan/ton.
7   For example, Kornai's (1980) *Economics of Shortage*.
8   Excepting coal for electricity generation, which was then still controlled.
9   Although the diseconomies of scale would have existed even without the entry of TVEs, the situation was actually much worse than before.
10 This is a common problem in China. In addition to coal, other products such as steel, cement, coke and electricity are all in surplus. Some of the relevant departments set a minimum price in order to protect the sector in 2000, an action criticized by some scholars.
11 Recovery rate is the proportion of a deposit successfully removed by the miners.

12 In 1995 there were 14,432 TVE coalmines extracting coal in key SOE coalfields, of which 7,526 or 52 per cent had no licences at all (Ye and Zhang 1998: 83).
13 The vice-director of China National Coal Association admitted that 'it is [TVE mines'] output rather than numbers that told the truth' (interviewed on 4 July 2003 in Beijing).
14 Yujialiang mine was expected to produce 4 Mt of blending coal per year for Shenhua. Since the lease fee was 10 yuan per ton, Dianta Township would receive at least 40 million yuan per year, compared with the annual 5 million yuan revenue. Besides, Shenmu County would receive 12.5 per cent of national tax paid by Yujialiang mine.

4 Transition: transforming traditional SOE coalmines

1 The Jixi Mining Bureau was renamed as the Jixi Mining Group in 2001.
2 It was, on average, 15 per cent in Shanxi province before 1985.
3 It was, on average, over 60 per cent in Shanxi province by 1994.
4 A triangular debt is the situation in which a series of companies fall into a debt chain. For example, company A is owed by company B, but company B cannot repay its debt to A because it is waiting to be paid by company C. Thus, B may ask C to pay company A directly.
5 The accident happened on 20 June 2002 in Chengzihe Mine of Jixi Mining Bureau, in which 129 were killed, including its general manger, Mr Zhao, and others who had greatly assisted the author's research at Jixi. It was admitted that the accident was caused by a large gas explosion resulting from neglect. Chengzihe Mine had 5,500 employees producing 1.1 Mt of coal in 2002 (Xinhua News, 23 June 2002).
6 For example, customers of Jixi paid their debts to Jixi with products, commodities or materials rather than cash. It should be noted that material rather than cash payments are popular in China's coal sector, and also quite popular in some other sectors. People jokingly said that China had returned to a barter society again.
7 Redundancies started in Jixi during the 1980s, but larger scale job losses occurred after 1994. Jixi had over 100,000 employees before 1990.
8 Historically a coalworker's wife rarely worked because she had to look after an overworked husband and because of the lack of suitable jobs for women in coalmines.
9 The demonstrators (mainly pensioners) were forced by the Ministry to go to Heilongjiang in a special carriage of the train. Ticket were paid for by the Ministry, but were later deducted from the demonstrators' pensions.
10 Because the coal industry is an underground industry, its production needs ventilation, roof support, gas and water exclusion, ground pressure stabilization, and other specialized technologies.
11 The director of Jixi, Mr Ren, said in 2000 that 'apart from a cemetery, Jixi Bureau has everything, just like a comprehensive welfare society. By 1995 Jixi had 106 schools, childcare and special education of all sorts, 2,465 hospital beds, over 100 kitchens and restaurants, dozens of local police stations' (interview at Jixi Mining Bureau Conference Hall, Jixi, 16 January 2000).
12 This has also caused huge amounts of corruption. It is extremely complicated to access a railhead and the fees are high, especially in some coal rich areas such as Shanxi.
13 Reflected by the market since 1980, a coal storage ratio (the ratio of stockpiled coal to total output) of over 13 per cent may imply the coal market is already in surplus. From 1990 to 1994, the storage ratios were 17.9 per cent, 19 per cent, 19.6 per cent, 18 per cent and 17.2 per cent respectively, which indicated that China's coal market might have been in surplus from 1990 (CERC 1995: 36).
14 Lacking market information, the government was still encouraging TVE coalmines to produce more, and was still predicting the coal demand would be increased to over 1.4 billion tons by 2000.

15 Consequently the total amount of tax payment was increased. During the 33 years (1953–85) from China's First Five-year Plan to the Sixth Five-year Plan, the average taxes were 458 million yuan each year, which was increased to 993 million yuan each year during the Seventh Five-year Plan (1986–90) (CCEC 1999: 562).

16 From 1985 to 1993, there was a 35.20 billion yuan increase in payments to banks by key SOE coal bureaux due to this investment policy change. As the government's total subsidy during this time was 28.66 billion yuan, SOE coal bureaux had to use their own funds of some 20 billion yuan to pay the loan and interests to banks (CERC 1995: 193).

17 Before the government announced its decision to stop subsidies in 1993, Jixi received some 300 million yuan each year.

18 To be fair, insufficient prepared mining capacity has been an ongoing problem, not only in Jixi but also in other SOEs, not only in the coal sector, but also in other sectors such as aluminium, copper and especially in the iron-ore industries. Between 1952 and 1978, the average annual increase in coal production was 1.3 times higher than the average annual increase in capacity. In 1977 and 1978, the increases in production were respectively 4.8 and 5.8 times the increase in capacity, and the situation had reached crisis proportions. The average annual increase in capacity between 1985 and 1989 was 24 million tons, while the average annual increase in production was over twice this, at 53 million tons (Thomson 1996: 735).

19 For example, in 1989 the coal-processing industry, other upstream industries and the transport sector received a 3.3 billion yuan transfer because of the low price of coal (Albouy 1991: 10–12), equivalent to 66 per cent of total losses in state coalmines in the same year (SSB 1990: 442, quoted in Wang 1998: 58).

20 These coalmines share a certain amount of subsidies from Jixi, received annually from the central government. Now that they are going to be bankrupted, their shared subsidies would be used to pay part of their bankruptcy costs.

21 'Market transition is a conflict-ridden rather than a consensual process' (UNDP 2000: 7).

5 Globalization: building competitive coal corporations

1 'Blending coal' means mixing different coal types to arrive at a specific coal product which exactly meets the customer's requirements on some specified index such as calorie and sulphur content.

2 According to China's current administrative structure, the coal resource beneath the Shenfu Coalfield is administered by the Ministry of Land and Resources, the railway by the Ministry of Railways and the port by the Ministry of Communication. Consequently the construction at Shenhua required approval from each of these ministries. In addition, Shenfu Coalfield is located in both Shaanxi province and Inner Mongolia Autonomous Region, the railway must cross each of Shaanxi, Shanxi and Hebei provinces, and the port is located in Hebei province, Shenhua's construction therefore also needed the full cooperation of and local infrastructure support from local governments.

3 The effects were that, after starting to use its own railway and port in 2001, Shenhua reduced its coal cost by more than 40 yuan per ton. In addition, Shenhua is expected to repay all the loans and interest by 2015, so that an additional 40 yuan per ton can be removed from current interest and loan payments. Thus, the total cost of Shenhua coal (including freight) can be brought below 120 yuan per ton (about $15) by 2015, which would obviously greatly improve its competitive capacity in the international market.

4 By 2002, China's domestic savings had reached 18,402 billion yuan ($2,244 billion), of which Chinese residents' savings were 8,691 billion yuan ($1,060 billion) (CSD 2003: 84).

5 Energy security requires an affordable, reliable and sustainable supply of energy for commercial and household use. Low-cost and uninterrupted energy supply is fundamental for economic development and political stability. China's emergence as a major world industrial producer means that its energy security is now of international concern. Nolan *et al.* 2004 present the first in-depth case study of China's pioneering coal liquefaction project carried out in the Shenhua Group. This project is of great potential significance for China's energy security. It is likely to have important implications for China's position within the structure of global energy supply, and have wide potential implications for international business and relations.

6 Which amounts to 16 billion yuan ($1.9 billion). After making innovations to the direct coal liquefaction technology originated from the American HTI Company, Shenhua started to construct the coal liquefaction production plant in 2003. The first step of the programme aims to produce 1 Mt of oil (petrol, diesel, liquid gas and other products) by 2005 using 7 Mt of unwashed coal. The investment in this phase will be 7.7 billion yuan. If this first production line is successful, the project will be expanded to 5 Mt oil products by 2010 and through to 2020, while at the same time, another line using direct coal liquefaction technology and another using indirect coal liquefaction technology also start construction, enabling Shenhua to have a total of 30 Mt of oil products by 2020. Although there are about 27 proposals from over 13 provinces which intend to start coal liquefaction production, the Chinese central government has only approved Shenhua to take action (author's interview conducted in China in June and July 2003).

7 While investing in mine construction would just have produced more unsold coal which was already in surplus.

8 Shenhua planned to produce 4 Mt of coal from the two mines, at a rental fee of 8 yuan per ton. However, after merger and innovations, the two coalmines now produce a total of 16 Mt per year.

9 No single mining company was included in 'Ranking of top 300 international companies by R&D investment within industry' (DTI 2000: 46–59). In 1999 Rio Tinto spent $61 million on research (and $149 million for exploration and evaluation) (Rio Tinto 1999 Annual Report and Accounts: 51). BHP spent $94 million in 1999–2000, $221 million in 1998–9 and $174 million in 1997–8 on research and development (BHP Financial Statements 2000: 59).

10 Three of the FWB coal bureaux were making huge losses and one was under construction at the time. The subsidies requested from the central government by Inner Mongolia as a condition of accepting the FWB were unacceptably high. As a wholly state-owned enterprise, Shenhua had no alternative but to accept the government's decision, but since then Shenhua has implemented a two 'no's policy to the FWB so that they still remain as independent accounting units: 'no obligation to help them repay their loans and liabilities' and 'no guarantees provided for them for loans from banks'. However, Shenhua admits that sometimes it has had to help them in various ways. The situation is far from resolved, and is potentially a huge constraint on Shenhua's competitive position. In 2001 Shenhua spent 140 million yuan to help three of the Five Western Bureaux upgrade their production techniques. By 2003 some positive signs emerged as some of bureaux had started to make profits.

11 Of these, 9,000 are in the coal business, and a further 13,000 are employed in Shenhua's power plants, Baoshen Rail Ltd, Shuohuang Rail Ltd, Huanghua Port Ltd, and some other subsidiary companies.

12 This is the case not only in the coal sector; other sectors such as steel, cement and power stations all have the same problem of surpluses resulting from the rapid growth in output from TVE producers.

13 For example, Shenhua as a state-owned large company has not only to provide school and accommodation for its own employees and their families, but also to build schools, roads, environmental facilities and so on for local communities.

14 On 16 July 2003, thirteen days after the author's second interview with Ye Qing, the Department of Organization of the Chinese Communist Party announced that Ye was stepping down from the position of Chairman and Party Secretary of Shenhua. The current president Chen Biting was nominated as his successor. It was emphasized that this is a normal leadership change, based on the retirement age regulations for SOE top managers (Shenhua Group, 17 July 2003).
15 Big business worldwide is entering a period of great difficulty due to the global economic slowdown, which may offer large Chinese firms new opportunities to catch up.

6 Conclusion: inter-relationship of the three challenges and the role of the state

1 The state is a convenient term for government, although in China's context, the state tends to be the nexus of political power consisting of the Communist Party and the organs of government.
2 The name was then Huaneng Clean Coal Corporation.
3 Which was reflected in the similar proposals that 'the state builds rail while individuals open mines', and 'the state, collectives and individuals cooperate together to develop Shenfu Coalfield', which were repeated by Hu Yaobang and Li Peng when they visited the Shenfu Coalfield in 1985 and Zhao Ziyang when he visited it in 1986. Zhao even innovatively proposed to establish Huaneng Clean Coal Corporation to be a shareholding company with public and private funding. This example clearly reflects the embarrassing situation where the government desperately wanted more coal, but worried that the government funds were not sufficient to build more coalmines.
4 Besides, all top leaders visited Shenhua's coalfield in its early development phase, so that they could understand the coalfield possessed all the necessary conditions to build a world-class modern large coal corporation: such as very shallow depth of coal seams, excellent geological condition, low-gas, low-sulphur and high calorific coal.
5 This book uses the definition given by Chang (1996: 60), which defines industrial policy as a policy aimed at particular industries to achieve the outcomes that are perceived by the state to be efficient for the economy as a whole. According to this definition, the Chinese government did implement a lot of industrial policies for the coal industry, but its purpose was to raise efficiency for the economy as a whole.
6 To understand the distinction between royalty and rent in a global context, see Fine 1990: 50.
7 For example, China's most important laws and regulations on coal listed above were exactly to solve the problems emerging from the fast development of China's coal industry during the last two decades.
8 In 2000 the division of rail carriage was 42 per cent coal, 20 per cent steel and metallurgical goods, etc, 6 per cent construction materials, 6 per cent agricultural goods and 5 per cent oil.
9 This is a very powerful corporation in the Shanxi coal industry. Between 1985 and 1996 it was responsible for transporting and selling all Shanxi's coal except that from a few companies such as Datong. It had 12 transport-and-sale subsidiaries at prefecture and county level. It controlled 10 coal railheads (*jimeizhan*) along with railways in Shanxi. Most non-SOE coal could not be sold without their cooperation.
10 Including an energy base fund of 20 yuan, a supplementary production fee of 10 yuan, a special maintenance fee of 5 yuan, a service fee of 2.5 yuan and a water resource supplement fee of 2 yuan.

11 Including a depletion fee of 1 per cent, a management fee of 3 per cent, a service fee of 2.5 per cent, and a profit share of 5 per cent.

12 By 1995 a normal coal transport train could carry 2,500 to 2,700 tons and a heavy train 3,500 tons, compared to over 10,000 tons per train in USA. The highest speed was 100 km per hour in China (CERC 1995: 72).

13 As in Shanxi, when numerous middlemen operated railheads and were involved in coal transport and sale, the market was chaotic. Later when Shanxi Coal Transport-Sale Corporation centralized all activities of coal transport and sale in Shanxi, monopoly problems occurred.

14 Normal profits earned from running railways are handed over to the state, but these 'extra fees' can be retained by the rail bureaux. It is, therefore, common for railway companies, although publicly owned, to make profits for themselves through changing the way services are named and charged for.

15 The current capacity for all SOE coalmines is about 600 Mt, and this may reach 800 Mt by 2010, based on an estimated 20 Mt increase in capacity each year.

16 Central government only supplements the difference between the total cost of the closing and bankruptcy and the amount firms can provide from the sale of their assets and from other sources.

17 The former SETC issued several emergency documents to key SOE coal companies and relevant departments to deal with the 'shortage' problems. However, deeper research found that there was not a real shortage in 2002, as the aggregate coal output should be able to meet the aggregate demand. Instead, the real problem was that coal could not be transported from production areas to demand areas in time and in adequate amounts. This information once again indicates the importance of the existence of a certain number of TVE mines, especially in some particular areas.

18 When I first made the point and figure, I had not realized that Dobb (1967: 88) made a similar point and figure forty years earlier. I am very grateful that my point was initiated by a talk with my supervisor Peter Nolan, who even drew a draft figure for me to illustrate his point, although this was neither the same as Dobb's figure nor the one shown here.

References

Adelman, I. (2001) 'Fallacies in development theory and their implications for policies', in Meier, M.G. and Stiglitz, E.J. (eds) *The Frontier of Development Economics*, New York: Oxford University Press.

Albouy, Y. (1991) *Coal Pricing in China: Issues and Reform Strategy*, World Bank Discussion Papers, No. 138. Washington, DC: World Bank.

Amsden, A.H. (1989) *Asia's Next Giant: South Korea and Late Industrialization*, Oxford: Oxford University Press.

Amsden, A.H., Kochanowicz, J. and Taylor, L. (1994) *The Market Meets Its Match: Restructuring the Economics of Eastern Europe*, Cambridge, MA: Harvard University Press.

Asian American Coal (2001) Speech at the Asia/China Coal Market and Technology Conference, 18 September 2001, Beijing.

Åslund, A. (1991) *Post-Communist Economic Revolutions: How Big a Bang?*, New York: Cambridge University Press.

—— (2002) *Building Capitalism: The Transition of the Economic Soviet Bloc*, Cambridge: Cambridge University Press.

Banister, J. (1987) *China's Changing Population*, Stanford: Stanford University Press.

Bell, W.B., Khor, H. and Kochhar, K. (1993) *China at the Threshold of a Market Economy*, Occasional Paper No. 107. Washington, DC: International Monetary Fund.

Bhagwati, J. (1985) *Essays on Development Economics*, Vol. 1, Cambridge, MA: MIT Press.

Bhagwati, J. and Krueger, A. (1973) 'Exchange control, liberalization, and economic development', *American Economic Review*, 63: 420–7.

BHP Billiton (2003) *BHP Billiton Annual Report*. Available online. http://www.bhpbilliton.com/bbContentRepository/Reports/LimitedAnnualReport.pdf, (accessed 30 April 2003).

Blanchard, O.J. and Layard, R. (1991) *How to Privatize*, Discussion paper no. 50, Centre for Economic Performance, London School of Economics and Political Science, London.

Bramall, C. (1989) *Living Standards in Sichuan, 1931–1978*, London: Contemporary China Institute, School of Oriental and African Studies, University of London.

—— (1993) *In Praise of Maoist Economic Planning: Living Standards and Economic Development in Sichuan Since 1931*, Oxford: Clarendon Press.

Breth, R.M. (1977) *Mao's China: A Study of Socialist Economic Development*, Melbourne: Longman Cheshire.

Buck, J.L. (1937) *Land Utilization in China: A Study of 16,786 Farms in 168 Localities, and 38,256 Farm Families in Twenty-two Provinces in China, 1929–1933*, Oxford: Oxford University Press.

Business Week Online (2002)'China's Big Bet on Gas: A new pipeline may wean the country off coal and oil', 29 April 2002. Available online. http://www.businessweek.com/magazine/content/02_17/b3780129.htm (accessed 29 May 2002).

Byrd, W.A. and Lin, Q.S. (eds) (1990) *China's Rural Industry, Structure, Development, and Reform*, Oxford: Oxford University Press.

Cao, Y., Fan, G. and Woo, W.T. (1997) 'Chinese economic reform: past success and future challenges', in Woo, W.T., Parker, S. and Sachs, J. (eds) *Economies in Transition: Comparing Asia and Eastern Europe*, Cambridge, MA: MIT Press.

CCC (China Coal Consultancy) (2000) 'Statement on the reform and development of China's coal industry' [*zhongguo meitan gongye gaige yu fazhan zongshu*], memo, Beijing: CCC. In Chinese.

CCCIEC (Contemporary China's Coal Industry Editorial Commission) (1989) *Contemporary China's Coal Industry* [*dangdai zhongguo de meitan gongye*], Beijing: China's Social Science Press. In Chinese.

CCEC (China Coal Editorial Commission) (1999) *History Record of China's Coal, Comprehensive Volume* [*zhongguo meitan zhi, zonghe juan*], Beijing: Coal Industry Publishing House. In Chinese.

CCIY (China's Energy Development Report Editorial Committee) (1999) *China Coal Industry Yearbook 1999* [*zhongguo meitan gongye nianjian*] Beijing: Coal Industry Publishing House. In Chinese.

CEDREC (China Energy Development Report Editorial Committee) (2001) *The Report of China's Energy Development* [*zhongguo nengyuan fazhan baogao*], Beijing: China Computing Publishing House. In Chinese.

CEPY (China Electric Power Yearbook Editorial Committee) (1998) *China Electric Power Yearbook* [*zhongguo dianli nianjian*], Beijing: China Electric Power Publishing House. In Chinese.

—— (2002) *China Electric Power Yearbook* [*zhongguo dianli nianjian*], Beijing: China Electric Power Publishing House. In Chinese.

CERC (China Energy Research Committee) (1995) *Research Report on Strategic Choices to Establish Market Economic System in China's Coal Industry* [*zhongguo meitan gongye chuangjian shichang jingji tizhi de zhanlu xuanze*], Beijing: Ministry of Coal Industry. In Chinese.

CESRY (2001) *China Economic System Reform Yearbook* [*zhongguo jingji tizhi gaige nianjian*], Beijing: CESRY Publishing House. In Chinese.

CEY (China Economic Yearbook) (1998) *China Economic Yearbook* [*zhongguo jingji nianjian*], Beijing: China Economic Yearbook Publishing House. In Chinese.

—— (2000) *China Economic Yearbook* [*zhongguo jingji nianjian*], Beijing: China Economic Yearbook Publishing House. In Chinese.

—— (2002) *China Economic Yearbook* [*zhongguo jingji nianjian*], Beijing: China Economic Yearbook Publishing House. In Chinese.

Chai, C.H. (1997) *China: Transition to a Market Economy*, Oxford: Clarendon Press.

Chandler, A.D. (1990) *Scale and Scope: The Dynamics of Industrial Capitalism*, Cambridge, MA: Belknap Press of Harvard University Press.

Chang, H.J. (1996) *The Political Economy of Industrial Policy*, Basingstoke: Macmillan.

—— (2002) *Kicking Away the Ladder?: Economic Development in Historical Perspective*, London: Anthem.

Chang, Y.S. and Lo, D. (2002) 'Explaining the financial performance of China's industrial enterprises: beyond the competition-ownership controversy', *China Quarterly*, June, 170: 413–40.

Chang, H.J. and Singh, A. (1992) 'Public enterprises in developing countries and economic efficiency: a critical examination of analytical, empirical, and policy issue', *UNCTAD Review*, No. 3, 45–82. Geneva: United Nations Conference on Trade and Development.

Chen, J.Y. (1989) 'China's transfer of the surplus agricultural labor force (TSALF)', in Longworth, J.W. (ed.) *China's Rural Development Miracle with International Comparisons*, Brisbane: University of Queensland Press.

Chen, B. (2003) 'Report for 2003 working conference of the Shenhua Group', Available online. http://www.shenhuagroup.com.cn/ReadNews.asp?NewsID=185 (accessed 14 Jan 2003). In Chinese.

Chen, K., Jefferson, G H. and Singh, I. (1992) 'Lessons from China's economic reform', *Journal of Comparative Economics*, 16(2): 201–25.

Chen, J., Thomas, C.S. and Liao, M. (2001) 'Pension reform blues', *The China Business Review*, (May/June) 3: 18.

Chinaed (2003) 'Economist forum'. Available online. http://www.chinaed.com.cn:7777/ Detail.wct?RecID=178&SelectID=2&ChannelID=4744 (accessed 15 May 2003). In Chinese.

Clark, D. (1999) 'Conceptualizing development', unpublished thesis, University of Cambridge.

CNCIEC (China Coal Import and Export Corporation) (2002) 'China's coal export', paper presented at Coaltrans China 2002 Conference, September 2002, Beijing.

Coaltrans China (2004) Annual international conference on China's coal industry, 22–23 March 2004, Shanghai.

Croll, E. J. and Huang, Ping (1996) 'Migration for and against agriculture', Research report, SOAS (School of Oriental and Africa Studies), University of London.

CSD (China Statistical Digest Editorial Committee) (2000) *China's Statistical Digest*. Beijing: China Statistical Press.

—— (2002) *China's Statistical Digest*, Beijing: China Statistical Press.

—— (2003) *China's Statistical Digest*, Beijing: China Statistical Press.

CSSA (China's Social Science Academy) (2000) *China's Industrial Development Report 2000*, Beijing: China's Social Science Academy Press.

—— (2001) *China's Industrial Development Report 2001*, Beijing: China's Social Science Academy Press.

——(2002) *China's Industrial Development Report 2002*, Beijing: China's Social Science Academy Press.

CTVEYEC (China Township and Village Enterprises Yearbook Editorial Committee) (2000) *China's Township and Village Enterprises Yearbook* [*xiangzhen qiye nianjian*], Beijing: China Agriculture Press.

Daliuta Mine (2000) 'To pursue high productivity and high efficiency – the construction of Daliuta' [*zhuiqiu gaochan gaoxiao – jianshe Daliuta*], Memo, Daliuta Mine, Shenhua Group. In Chinese.

Deng, Xiaoping (1984 [1994]) 'Our great goal and fundamental policy', in *Selected Works of Deng Xiaoping, 1982–1992*, Beijing: Foreign Languages Press.

—— (1992 [1994]) 'Key points of speeches in Wuchang, Shenzhen, Zhuhai and Shanghai', in *Selected Works of Deng Xiaoping, 1982–1992*, Beijing: Foreign Languages Press.

Dobb, M. (1967) *Papers on Capitalism, Development and Planning*, New York: International Publishers.

DOE (Department of Energy, USA) (1996) 'China's energy: a forecast to 2015'. Available online: http://www.doe.gov (accessed 20 Sep 1999).

DTI (Department of Trade and Industry, UK) (2000). *The 2000 R&D Scoreboard, Company Data*, London: DTI.

Du, Rensheng (1989) 'Advancing amidst reform', in Longworth, J.W. (ed.) *China's Rural Development Miracle with International Comparisons*, Brisbane: University of Queensland Press.

Eckaus, R.S. (1955) 'Factor proportions in underdeveloped countries', *American Economic Review*, vol. 45, September.

Ellman, M. (2000a). 'The Russian economy under Yeltsin', *Europe-Asia Studies*, 52(8): 1417–32.

—— (2000b) 'The social costs and consequences of the transformation process', in United Nations (ed.), *Economic Survey of Europe 2000*, 2/3, Geneva: United Nations.

Fan, G. and Woo, W.T. (1996) 'State enterprise reform as a source of macroeconomic instability: the case of China', *Asian Economic Journal*, 10(3): 207–24.

FAO (1999) *Production Yearbook*, vol 53. Rome: United Nations Food and Agricultural Organization.

Fei, J.C.H. and Ranis, G. (1963) 'Capital accumulation and economic development', *American Economic Review*, June 53: 283–313.

—— (1964) *Development of the Labour Surplus Economy: Theory and Policy*, Irwin: Homewood.

Fine, B. (1990) *The Coal Question: Political Economy and Industrial Change from the Nineteenth Century to the Present Day*, London: Routledge.

FT (*Financial Times*) (2001) 'Long romance for a union of substance', 24/25 March 2001: 14.

Ge, Fu (1991) 'On changing dual-track coal price system to one track system', in Ministry of Energy (ed.) *Paper Collection on Energy Policy Research* [*nengyuan zhengce yanjiu wenji*], 1: 199–210, Beijing: Ministry of Energy of PRC. In Chinese.

Ghatak, S. (1978) *Development Economics*, London: Longman.

Goldman, M I. (1996) *Lost Opportunity: What has Made Economic Reform in Russia So Difficult?*, 2nd edn, New York: Norton.

Gore, C. (2000) 'The rise and fall of Washington Consensus', *World Development*, 28(5): 789–804.

Griffin, K. (1996) *Studies in Globalization and Economic Transition*, London: Macmillan.

Hassard, J., Sheehan, J. and Morris, J. (1999) 'Enterprise reform in post-Deng China', *International Studies of Management and Organization*, 29(3): 54–83.

Held, D., Anthony, G., McGrew, D., Goldblatt, and Perraton, J. (1999) *Global Transformations: Politics, Economics and Culture*, Cambridge: Polity Press.

Ho, S. (1994) *Rural China in Transition: Non-agricultural Development in Rural Jiangsu, 1978–1990*, Oxford: Oxford University Press.

—— (1995) 'Rural non-agricultural development in post-reform China: growth, development and issues', *Pacific Affairs*, 68(3): 360–91.

Hogendorn, J. S. (1996) *Economic Development*, New York: HarperCollins College Publishers.

Horii, N. (2001) 'Coal industry: development of small coal mines in market transition and its externality', in Horii, N. and Gu, Shuhua (eds) *Transformation of China's Energy Industries in Market Transition and Its Prospects*, Chiba, Japan: Institute of Developing Economies, Japan External Trade Organisation.

Horii, N. and Gu, Shuhua (eds) (2001) *Transformation of China's Energy Industries in Market Transition and Its Prospects*, Chiba, Japan: Institute of Developing Economies, Japan External Trade Organisation.

Huang, C.C.P. (1985) *The Peasant Economy and Social Change in North China*, Stanford: Stanford University Press.

Huntington, S.P. (1968) *Political Order in Changing Societies*, New Haven: Yale University Press.

Hussain, A. (1983) 'Economic reforms in Eastern Europe and their relevance to China', in Feuchtwang, S. and Hussain, A. (eds) *The Chinese Economic Reforms*. New York: St Martin's Press.

IMF (International Monetary Fund) (2000) *World Economic Outlook: Focus on Transition Economies*, Washington, DC: IMF Publication Service.

IMF (International Monetary Fund), World Bank, OECD and EBRD (1990) *The Economy of the USSR: Summary and Recommendations*, Washington, DC: World Bank.

Ishikawa, S. (1967) *Economic Development in Asian Perspective*, Tokyo: Kinokuniya Bookstore.

Jefferson, G., Hu, A., Guan, X. and Yu, X. (2003) 'Ownership, performance, and innovation in China's large- and medium-size industrial enterprise sector', *China Economic Review*, 14(1): 89–113.

Jin, H. and Qian, Y.Y. (1998) 'Public versus private ownership of firms: evidence from rural China', *The Quarterly Journal of Economics*, August, 113(3): 773–809.

Jixi Mining Bureau (1985) *Statistical Bulletin*, Jixi: Jixi Mining Bureau.

—— (1988) *Statistical Bulletin*, Jixi: Jixi Mining Bureau.

—— (1991) *Statistical Bulletin*, Jixi: Jixi Mining Bureau.

—— (1994) *Statistical Bulletin*, Jixi: Jixi Mining Bureau.

—— (1998) *Statistical Bulletin*, Jixi: Jixi Mining Bureau.

—— (2000) *Statistical Bulletin*, Jixi: Jixi Mining Bureau.

Killick, T. (1989) *A Reaction Too Far: Economic Theory and the Role of the State in Developing Countries*, London: Overseas Development Institute.

Kolodko, G.W. (1998) *Ten Years of Postsocialist Transition: The Lessons for Policy Reforms*, Washington, DC: Development Economics Research Group, The World Bank.

—— (2000) *From Shock to Therapy: The Political Economy of Postsocialist Transformation*, Oxford: Oxford University Press.

Kornai, J. (1980) *Economics of Shortage*, vols. A and B, Amsterdam: North Holland.

—— (1990) *The Road to a Free Economy: Shifting from a Socialist System – The Example of Hungary*, New York: W.W. Norton.

—— (1992a) 'The post-socialist transition and the state: reflection in the light of Hungarian fiscal problems', *The American Economic Review*, 82(2): 1–21.

—— (1992b) *The Socialist System: The Political Economy of Communism*, Oxford: Clarendon Press.

—— (1994) 'Transformational recession: the main causes', *Journal of Comparative Economics*, 31(1–2): 125–36.

—— (2000) 'Making the transition to private ownership', *Finance and Development*, 37(3): 1–4.

Krueger, A. (1974) 'The political economy of the rent-seeking society', *American Economic Review*, June, 64(3): 291–302.

—— (1979) *The Development Role of the Foreign Sector and Aid*, Cambridge, MA: Harvard University Press.

—— (1983) *Trade and Employment in Developing Countries: Synthesis and Conclusions*, Chicago: University of Chicago Press.

Lau, W.K. (1999) 'The 15th congress of the Chinese communist party: milestone in China's privatisation', *Capital and Class*, Summer, Issue 68: 51–87.

Lefeber, L. (2000) 'Classical vs. neoclassical economic thought in historical perspective: the interpretation of processes of economic growth and development', *History of Political Thought*, XXI(3): 525–42.

Lewis, A. (1954) 'Economic development with unlimited supplies of labour', in Corbridge, S. (1999) (ed.) *Development: Critical Concepts in the Social Sciences*, vol. 3. London: Routledge.

—— (1979) 'Development strategy in a limping world economy', The Elmhurst Lecture, The International Conference of Agricultural Economists, 3–12 September, Banff, Canada.

Lin, Justin Yifu and Yao, Yang (1999) 'Chinese rural industrialization in the context of the east Asian miracle', unpublished paper at China Center for Economic Research, Beijing University.

Luo, Y., Tan, J.J. and Shenkar, O. (1998) 'Strategic responses to competitive pressure: the case of township and village enterprises in China', *Asia Pacific Journal of Management*, 15(1): 33–50.

—— (1999) 'Township and village enterprises in China: strategy and environment', in Kelley, M. and Luo, Y. (eds) *China 2000, Emerging Business Issues*, London: Sage Publications Inc.

Ma, Hong and Wang, Mengkui (eds) (2001) *China Development Studies: The Selected Research Reports of DRC of the State Council* [*zhongguo fazhan yanjiu*], Beijing: China Development Press. In Chinese.

Marx, K. (1887 [1961]) *Capital: A Critical Analysis of Capitalist Production*, Moscow: Foreign Languages Publishing House.

Marx, K. and Engels, F. (1888 [1967]) *The Communist Manifesto*, Harmondsworth: Penguin Books.

MCI (Ministry of Coal Industry) (1996) *The Prospect on Coal Demand and Supply*, [*meitan xuqiu yu gongyin zhanwang*], Beijing: MCI. In Chinese.

Meng, Xin (2000) *Labour Market Reform in China*, New York: Cambridge University Press.

Mergers and Acquisitions (2001) '10-Year global merger completion record 1991–2000', *Mergers and Acquisitions*, 36(2): 23, 39.

Ministry of Energy (ed.) (1991, 1992) *Collection of Papers on Energy Policy Research* [*nengyuan zhengce yanjiu wenji*], vols 1 and 2. Beijing: Ministry of Energy. In Chinese.

Myint, H. (2001) 'International trade and the domestic institutional framework', in Meier, M.G. and Stiglitz, E.J. (eds) *The Frontier of Development Economics*, New York: Oxford University Press.

Naughton, B. (1995) *Growing Out of the Plan: Chinese Economic Reform, 1978–1993*, Cambridge: Cambridge University Press.

Nolan, P. (1988) *The Political Economy of Collective Farms: An Analysis of China's Post-Mao Rural Economic Reforms*, Cambridge: Polity Press.

—— (1993) *State and Market in the Chinese Economy: Essays on Controversial Issues*, Basingstoke: Macmillan.

—— (1994) 'Introduction, the Chinese puzzle', in Fan, Q.M. and Nolan, P. (eds) *China's Economic Reforms: The Cost and Benefits of Incrementalism*, New York: St Martin's Press.

—— (1995) *China's Rise, Russia's Fall: Politics and Economics in the Transition from Stalinism*, Basingstoke: Macmillan.

—— (1999) 'Strategic choices in the development of the Chinese coal industry: the case of the Shenhua Group', unpublished paper at The Judge Institute of Management Studies, University of Cambridge.

—— (2001a) *China and the Global Business Revolution*, Basingstoke: Palgrave.

—— (2001b) *China and the Global Economy: National Champions, Industrial Policy and the Big Business Revolution*, Basingstoke: Palgrave.

—— (2003) *Transforming China*, Forthcoming.

Nolan, P. and Dong, Fureng (eds) (1989) *Market Forces in China: Competition and Small Business – The Wenzhou Debate*, London: Zed Books.

Nolan, P. and Wang, Xiaoqiang (1999) 'Beyond privatisation: institutional innovation and growth in China's large state-owned enterprises', *World Development*, 27(1): 169–200.

Nolan, P., Shipman, A. and Rui, H. (2004) 'Coal liquefaction, Shenhua Group, and China's energy security', *European Management Journal*, 2(2).

Nurkse, R. (1953) *Problems of Capital Formation in Underdeveloped Countries*, Oxford: Oxford University Press.

Osiris (2003) Company report. Available online. http://www.osiris.bvdep.com/cgi (accessed 5 May, 2003).

Outlook (2003) 'China faces huge pressure of employment'. Available online. http://www1.cei.gov.cn/hottopic/doc/zjzt2002101/200309050669.htm (accessed 26 Sep 2003). In Chinese.

Pack, H. (2000) 'Industrial policy: growth elixir or poison?', *The World Bank Research Observer*, Feb.

Peng, Zhikui (ed.) (1999) *The Developmental Strategy of Shanxi Coal Economy* [*Shanxi meitan jinji fazhan zhanlu*], Beijing: Coal Industry Publishing House. In Chinese.

Perkins, D. (1994) 'Completing China's move to the market', *Journal of Economic Perspectives*, Spring, 8: 23–34.

Polanyi, K. (1957) *The Great Transformation: The Political and Economic Origins of our Time*, Boston: Gower Beacon Press.

Prosterman, R. (2001) 'Land tenure, food security and rural development in China', *Development: Local/Global Encounters*, 44(4): 79–84.

Prybyla, J.S. (1990) *Reform in China and Other Socialist Economies*, Washington, DC: AEI Press.

Pu, Hongjiu (1992) 'It is necessary to reform the price of coal-under-plan', in Ministry of Energy (ed.) *Collection of Papers on Energy Policy Research* [*nengyuan zhengce yanjiu wenji*], vol. 2. Beijing: Ministry of Energy. In Chinese.

Ranis, G. and Fei, J.C.H. (1961) 'A theory of economic development', *American Economic Review*, September, 51(4): 553–8.

Rawski, G.T. (1979) *Economic Growth and Employment in China*, Oxford: Oxford University.

Research Group (1997) *Research on China's Energy Strategy 2000 – 2050*, Beijing: China's Electric Power Press.

Ricardo, D. (1817 [1971]) *On the Principles of Political Economy and Taxation*, Harmondsworth: Penguin.

Rio Tinto China (2001) 'Foreign investment in China's coal industry', paper presented at the Asia/China Coal Market and Technology Conference, 17–18 September, Beijing.

Robinson, J. (1936) 'Disguised unemployment', *The Economic Journal*, June 46(182): 225–37.

Rosenstein-Rodan, P.N. (1943) 'Problems of industrialization of eastern and south-eastern Europe', *Economic Journal*, 53(210/211): 202–11.

Sachs, J. (1992a) 'Building a market economy in Poland', *Scientific American*, March: 20–6.

—— (1992b) 'The economic transition of Eastern Europe: the case of Poland', *Economics of Planning*, 25(1): 5–19.

—— (1992c) 'Privatization in Russia: some lessons from Eastern Europe', *American Economic Review*, 82(2): 43–8.

Sachs, J. and Woo, W.T. (1994) 'Structural factors in the economic reforms of China, Eastern Europe and the former Soviet Union', *Economic Policy*, 18: 102–45.

—— (1997) *Understanding China's Economic Performance*, NBER Working Paper No. 5935 Cambridge, MA: NBER.

Sachs, J., Woo, W.T. and Yang, X. (2000) 'Economic reforms and constitutional transition', *Annals of Economics and Finance*, 1(2): 423–79.

SCA (State Coal Administration) (1999a) *Tenth Five-year Plan and Long-term Plan by 2015 for China's Coal Industry* [*zhongguo meitan gongye dishige wunian jihua he 2015 nian changqi guihua*], Beijing: SCA. In Chinese.

—— (1999b) *Coal Economy Operation Report*, Beijing: SCA.

—— (2000) *Coal Economy Operation Report*, Beijing: SCA.

—— (2001) *Coal Economy Operation Report*, Beijing: SCA.

—— (2002a) *Coal Economy Operation Report*, Beijing: SCA.

—— (2002b) 'How to live without money: analysing why mine workers want money more than lives'. Available online. http://www.chinacoal.gov.cn/coal/jryw/020805x1.htm (accessed 1 Nov 2002). In Chinese.

—— (2002c) '16,000 small coalmines were closed nationwide'. Available online. http://www.chinacoal.gov.cn/coal/jryw/020925x1.htm (accessed 30 Dec 2002). In Chinese.

—— (2002d) 'Continued improvement of the coal industry'. Available online. http://www.chinacoal.gov.cn/coal/jryw/021211x2.htm (accessed 30 Dec 2002). In Chinese.

—— (2003a) *Coal Economy Operation Report*, Beijing: SCA.

—— (2003b) 'China's coal output is near 1.4 billion tons'. Available online. http://www.chinacoal.gov.cn/coal/jryw/030210x1.htm (accessed 10 Feb 2003). In Chinese.

SDRC (State Development Reform Commission) (2003) *Analysis Report of Industrial Demand for China's Coal* [*zhongguo meitan hangye xuqiu fenxi baogao*], Beijing: SDRC. In Chinese.

SETC (State Economic Trade Commission) (2001) 'Report on the performance of large industrial enterprise groups in 2000'. Available online. http://www.setc.gov.cn/gjzdqyxx/hyfx/200102260001.htm (accessed 26 Feb 2001). In Chinese.

—— (2002a) 'Continued growth of national key enterprises – the development report of the national key enterprises in 2001'. Available online. http://www.setc.gov.cn/gjzdqyxx/hyfx/200209130001.htm (accessed 13 Sep 2002). In Chinese.

—— (2002b) 'The Tenth Five-year Plan of the coal industry'. Available online http://www.setc.gov.cn/gjjmwznzn/gyswgh/200207310023.htm (accessed 20 Dec 2002).

—— (2003a) 'Report on the performance of key enterprises in major industrial sectors in 2002'. Available online. http://www.setc.gov.cn/gjzdqyxx/hyfx/200301270001.htm (accessed 27 Jan 2003). In Chinese.

—— (2003b) 'Performance of large industrial enterprise groups in coal, metallurgy, and construction material industries in 2002'. Available online. http://www.setc. gov.cn/gjzdqyxx/hyfx/200301290001.htm (accessed 29 Jan 2003). In Chinese.

Shendong Coal Ltd, Shenhua Group (1999) 'Statistical data of Shendong Coal Ltd from 1985–1998' [*shendong gongsi tongji ziliao 1985–1988*], memo. In Chinese.

—— (2003) 'Statistical data of Shendong Coal Ltd' [*shendong gongsi tongji ziliao*], memo. In Chinese.

Shi, Wanpeng (2002) 'Current situation of the coal industry'. Available online. http://www.chinacoal.gov.cn/coal/jryw/020131x1.htm (accessed 31 Jan 2002). In Chinese.

Shi, Qingqi and Zhao Jingce (eds) (1999) *China Industrial Development Report* [*zhongguo chanye fazhan baogao*], Beijing: China Zhigong Press. In Chinese.

Singh, A. (2002) 'Development economics', lecture presented at Faculty of Economics, University of Cambridge.

Smil, V. (1988) *Energy in China's Modernization: Advances and Limitations*, New York: M.E. Sharpe.

—— (1993) *China's Environmental Crisis: An Inquiry into the Limits of National Development*, New York: M.E. Sharpe.

—— (1998) 'China's energy and resource uses: continuity and change', *China Quarterly*, 156: 935–51.

Smith, A. (1776 [1863]) *An Inquiry into the Nature and Causes of the Wealth of Nations: A Concordance*, Edinburgh: A.&C. Black.

SPC (State Planning Commission) (1996) *Studies on Developmental Strategies of Large Enterprise Groups* [*daxing qiye jituan fazhan zhengce yanjiu*], Beijing: China's Economic Press. In Chinese.

—— (1997) *'97 Energy Report of China* [*97 baipishu: zhongguo nengyuan*], Beijing: China's Price Publishing House. In Chinese.

—— (1999a) *Studies on Developmental Model of Enterprise Groups* [*qiye jituan fazhan moshi yanjiu*], Beijing: SPC. In Chinese.

—— (1999b) *Studies on the Organizational Framework of Enterprise Groups* [*qiye jituan tizhi moshi yanjiu*], Beijing: SPC. In Chinese.

SSB (State Statistical Bureau) (1989) *China Energy Statistical Yearbook* [*zhongguo nengyuan tongji nianjian*]. Beijing: China Statistical Press. In Chinese.

—— (1985) *China Statistical Yearbook* [*zhongguo tongji nianjian*], Beijing: China Statistical Press. In Chinese.

—— (1995) *China Statistical Yearbook* [*zhongguo tongji nianjian*], Beijing: China Statistical Press. In Chinese.

—— (1996) *China Energy Statistical Yearbook* [*zhongguo nengyuan tongji nianjian*], Beijing: China Statistical Press. In Chinese.

—— (1998) *China Statistical Yearbook* [*zhongguo tongji nianjian*], Beijing: China Statistical Press.

—— (2000) *China Statistical Yearbook* [*zhongguo tongji nianjian*], Beijing: China Statistical Press.

—— (2001) *The Year 2000 Statistical Report of National Economy and Social Development* [*2000 nian guomin jingji he shehui fazhan tongji gongbao*], Beijing: China Statistical Press. In Chinese.

—— (2003) *China Statistical Yearbook* [*zhongguo tongji nianjian*], Beijing: China Statistical Press.

State Council (1991) 'The State Council endorses the State Planning Commission, the State Commission for Economic System Reform, and the Sate Council Production Office's request for permission to choose a batch of large enterprise groups to undergo trials' [*guowuyuan pizhuan guojia jiwei, guojia tigaiwei, guowuyuan shengchan bangongshi guanyu xuanze yipi daxing qiye jituan jinxing shidian qingshi de tongzhi*], State Publication, Number 71, issued on 24 December 1991. In Chinese.

—— (1997) 'The State Council endorses the State Planning Commission, State Economic and Trade Commission and the State Commission for Economic System Reform's opinions on deepening the trial work on large enterprise groups' [*guowuyuan pizhuan guojia jiwei, guojia jingmaowei, guojia tigaiwei guanyu Shenhua daxing qiye jituan shidian gongzuo de tongzhi*], State Publication, Number 15, issued on 29 April 1997. In Chinese.

State Environment Protection Bureau (2001). Speech at the Asian Coal Production/ Technology Conference, 17 September 2001, Beijing.

Steinfeld, E.S. (1998) *Forging Reform in China: The Fate of State-owned Industry*, Cambridge: Cambridge University Press.

Stiglitz, J.E. (1994) *Whither Socialism?*. Cambridge, MA: MIT Press.

—— (1999a) 'Whither reform? Ten years of transition', keynote address at the 1999 World Bank Annual Bank Conference on Development Economics, 28–30 April, Washington, DC.

—— (1999b) 'Quis custodiet ipsos custodes?', *Challenge*, November/December, 22(6): 26–67.

Sun, L. (2002) 'Fading out of local government ownership: recent ownership reform in China's township and village enterprises', *Economic Systems*, 26: 249–69.

—— (2000) 'The insider', *The New Republic*, April 17 and 24.

Sunday Times (2001) 'Billiton and BHP create a mining giant', 25 March 2001, 12.

Thomson, E. (1996) 'Reforming China's coal industry', *China Quarterly*, September, 147: 726–50.

—— (2003) *The Chinese Coal Industry: An Economic History,* London: Routledge Curzon.

UNCTAD (2000) 'A "New Deal" for development: the general summary from the United Nations Conference on Trade and Development, Tenth Session', *Presidents and Prime Ministers*, January–February: 10–18.

UNDP (United Nations Development Program) (2000) *China Human Development Report 1999: Transition and State*. Oxford: Oxford University Press.

Wade, R. (1990) *Governing the Market: Economic Theory and the Role of Government in East Asian Industrialization*, Princeton, NJ: Princeton University Press.

Wang, Xiaoqiang (1998) *China's Price and Enterprise Reform*, London: Macmillan.

Warriner, D. (1939) *Economics of Peasant Farming*. Oxford: Oxford University Press.

—— (1948) *Land and Poverty in the Middle East*, London: Royal Institute of International Affairs.

—— (1955) 'Land reform and economic development', Fifth Anniversary Commemoration Lectures, National Bank of Egypt, Cairo.

WCI (World Coal Institute) (2003) 'Coal data'. Available online. http://wci.rmid.co.uk/ uploads/CoalFacts03.pdf (accessed 5 Aug 2003).

Weitzman, M. and Xu, C. (1994) 'Chinese township-village enterprises as vaguely defined cooperatives', *Journal of Comparative Economics*, 18(2): 121–45.

Williamson, J. (1993) 'Democracy and the "Washington Consensus"', *World Development*, 21(8): 1329–36.

—— (1997) 'The Washington Consensus revisited', in Emmerij, L. (ed.) *Economic and Social Development into the XXI Century*, Washington DC: Inter-American Development Bank.

Wood, A. (1994) 'China's economic system: a brief description, with some suggestions for further reform', in Fan, Q.M. and Nolan, P. (eds) *China's Economic Reforms: The Cost and Benefits of Incrementalism*, New York: St Martin's Press.

World Bank (1995) *Bureaucrats in Business: The Economics and Politics of Government Ownership*, New York: Oxford University Press.

—— (1996) *World Development Report 1996: From Plan to Market*, New York: Oxford University Press.

—— (1997a) *Sharing Rising Incomes: Disparities in China*, Washington, DC: The World Bank.

—— (1997b) *Clear Water, Blue Skies*, Washington, DC: The World Bank.

—— (2002) *World Development Indicators*, Washington, DC: The World Bank.

World Resource Institute (2003) 'Earth Trends: the environmental information portal'. Available online. http://earthtrends.wri.org/text/ENG/variables/351.htm (accessed 15 Sep 2003).

Wu, Tiffany (2001a) 'WTO entry prods China to step up bank reform'. Available online. http://www.bbc.co.uk (accessed 19 September 2001).

Wu, Jinlian (2001b) *Reform: Now at a Critical Point* [*gaige: women zhengzai guo daguan*], Beijing: Living, Reading and New Knowledge – Three Combination Bookshop. In Chinese.

Xinhua News (2003) 'The updated news of the Jixi accident'. Available online. http://www.sina.com.cn (accessed 23 June 2002). In Chinese.

Xstrata (2003) 'Group information'. Available online. http://www.xstrata.com/prod_coal.php (accessed 23 April 2003).

Yan, Tianke, Zhang, Suo and Li, Debo (2000) 'The status of China's coal sector structure and the gap between that and of major coal producing countries' [*woguo meitan hangye jiegou xianzhuang jie yu zhuyao chanmei guojia de chaju*], *China Coal*, 4: 28–33. In Chinese.

Yang, Dali (1991) 'China adjusts to the world economy: the political economy of China's coastal development strategy', *Pacific Affairs*, Spring, 64(1): 42–64.

—— (2000) 'Economic development and poverty reduction in China over 20 years of reforms', *Economic Development and Cultural Change*, April 48(3): 447–74.

Yang, Jike (1997) 'Preface', in Yan, C.L. (ed.) *The Report on China's Energy Development* [*zhongguo nengyuan fazhan baogao*], Beijing: Economic Management Publishing House. In Chinese.

Ye, Qing (2003) 'Speech at 2003 working conference of the Shenhua Group'. Available online. http://www.shenhuagroup.com.cn/ReadNews.asp?NewsID=105 (accessed 16 Jan 2003). In Chinese.

Ye, Qing and Zhang, Baoming (eds) (1998) *China's TVE Coalmines*, Beijing: Coal Industrial Publishing House. In Chinese.

Yin, Ruiyu (2003) 'Achievements, key issues, and prospectives of China's steel industry', *Metallurgical Management*, 5: 4–11.

Zhang, Mei (2000) 'Internal migration and poverty', unpublished thesis, Faculty of Social and Political Sciences, University of Cambridge.

Zhang, Shuguang, Zhang, Yansheng and Wan, Zhongxin (1998) *Measuring the Costs of Protection in China*, Washington, DC: Institute for International Economics.

Zhou, Fengqi, and Zhou, Dadi (1999) *Study on Long Term Energy Development Strategies of China*, [*zhongguo zhongchangqi nengyuan zhanlu*], Beijing: China's Planning Press. In Chinese.

Zhou, Fenqi, Qu, Shiyuan, Han, Wenke and Su, Zhenming (1991) 'Strategies to relieve the stress of energy supply in our country', in Ministry of Energy (ed.) *Collection of Papers on Energy Policy Research* [*nengyuan zhengce yanjiu wenji*], Vol. 1. Beijing: Ministry of Energy. In Chinese.

Index

For Product Safety Concerns and Information please contact our EU
representative GPSR@taylorandfrancis.com
Taylor & Francis Verlag GmbH, Kaufingerstraße 24, 80331 München, Germany